AF775069

Monitoring of Marine Pollution

Edited by Houma Bachari Fouzia

Published in London, United Kingdom

IntechOpen

Supporting open minds since 2005

Monitoring of Marine Pollution
http://dx.doi.org/10.5772/intechopen.76739
Edited by Houma Bachari Fouzia

Contributors
Antigoni Zafirakou, Christopher Koutitas, Luis Soto, Alejandro Estradas-Romero, Diana Salcedo, Alfonso V. Botello, Guadalupe Ponce-Velez, Meng-Chuan Ong, Kamaruzzaman Yunus, Lucy Ngatia, Robert Taylor, Johnny Grace III, Daniel Moriasi, Eric Okuku, Veronica Wanjeri, Gilbert Owato, Maureen Kombo, Catherine Mwalugha, Nancy A. Oduor, Linet Kiteresi, Stephen Mwangi, Maria Virginia Alves Martins, Cintia Yamashita, Silvia Helena De Mello E Sousa, Eduardo Apostolos Machado Koutsoukos, Sibelle Trevisan Disaró, Jean-Pierre Debenay, Wânia Duleba, Sidrah Hafeez, Man Sing Wong, Sawaid Abbas, Houma Bachari Fouzia

© The Editor(s) and the Author(s) 2019
The rights of the editor(s) and the author(s) have been asserted in accordance with the Copyright, Designs and Patents Act 1988. All rights to the book as a whole are reserved by INTECHOPEN LIMITED. The book as a whole (compilation) cannot be reproduced, distributed or used for commercial or non-commercial purposes without INTECHOPEN LIMITED's written permission. Enquiries concerning the use of the book should be directed to INTECHOPEN LIMITED rights and permissions department (permissions@intechopen.com).
Violations are liable to prosecution under the governing Copyright Law.

Individual chapters of this publication are distributed under the terms of the Creative Commons Attribution 3.0 Unported License which permits commercial use, distribution and reproduction of the individual chapters, provided the original author(s) and source publication are appropriately acknowledged. If so indicated, certain images may not be included under the Creative Commons license. In such cases users will need to obtain permission from the license holder to reproduce the material. More details and guidelines concerning content reuse and adaptation can be found at http://www.intechopen.com/copyright-policy.html.

Notice
Statements and opinions expressed in the chapters are these of the individual contributors and not necessarily those of the editors or publisher. No responsibility is accepted for the accuracy of information contained in the published chapters. The publisher assumes no responsibility for any damage or injury to persons or property arising out of the use of any materials, instructions, methods or ideas contained in the book.

First published in London, United Kingdom, 2019 by IntechOpen
IntechOpen is the global imprint of INTECHOPEN LIMITED, registered in England and Wales, registration number: 11086078, The Shard, 25th floor, 32 London Bridge Street
London, SE19SG – United Kingdom
Printed in Croatia

British Library Cataloguing-in-Publication Data
A catalogue record for this book is available from the British Library

Additional hard and PDF copies can be obtained from orders@intechopen.com

Monitoring of Marine Pollution
Edited by Houma Bachari Fouzia
p. cm.
Print ISBN 978-1-83880-811-2
Online ISBN 978-1-83880-812-9
eBook (PDF) ISBN 978-1-83880-813-6

For EU product safety concerns:
IN TECH d.o.o., Prolaz Marije Krucifikse Kozulić 3, 51000 Rijeka, Croatia, info@intechopen.com or visit our website at intechopen.com.

We are IntechOpen, the world's leading publisher of Open Access books

Built by scientists, for scientists

4,200+
Open access books available

116,000+
International authors and editors

125M+
Downloads

151
Countries delivered to

Our authors are among the
Top 1%
most cited scientists

12.2%
Contributors from top 500 universities

BOOK CITATION INDEX
CLARIVATE ANALYTICS
INDEXED

WEB OF SCIENCE™

Selection of our books indexed in the Book Citation Index in Web of Science™ Core Collection (BKCI)

Interested in publishing with us?
Contact book.department@intechopen.com

Numbers displayed above are based on latest data collected.
For more information visit www.intechopen.com

Meet the editor

Houma Bachari Fouzia is a professor of oceanography, marine environment, and remote sensing. She has a PhD from the University of Science and Technology, Houari Boumediene, Algiers. She has a second doctoral degree in Sciences of the Universe and Environment, specializing in remote sensing and geography, from the University of Paris-Est Créteil, France. She has published 50 works in the field of monitoring marine pollution and remote sensing. She is the head of a research team, thesis supervisor of several PhD students working in various disciplines of marine science, such as ecotoxicology, impact studies, and remote sensing, and is the head of the research project "The Anthropogenic Impacts on the Marine and Coastal Environment." Prof. Fouzia is also involved in the monitoring of marine environment parameters, contributes to satellite measurements for the modeling of variables, and is a project manager integrating GIS satellite data and in situ measurements for monitoring the quality of coastal waters (modeling of ocean parameters). She has worked on 60 Projects related to graduations and masters on marine environmental science and coastal management. Prof. Fouzia was a member of the international project MerMex Marine Ecosystems Response in the Mediterranean Experiment 2010–2020 and is currently a member of the European research project Marine Database Management for Algeria and a member of the European research project Seadata, European Infrastructure of Oceanographic Data Management.

She is also the team leader of the "Ecotoxicology and Anthropogenic Impacts on the Marine Environment" Laboratory of Marine and Coastal Ecosystems and a director of the international journal *MarineScor*, Marine and Coastal Sciences Research.

Contents

Preface

Monitoring seawater and modeling ocean parameters require innovative methods to assess the impacts of marine pollution on the ecosystem. It is a decision-making process designed to evaluate environmental impacts and the effect of pollutants, which can be divided into long-term and short-term effects. Oil spill often has catastrophic effects on the marine biota when released. On the other hand, the study of the impact of human activities is necessary. We should care about the continuing environmental degradation of our oceans and coastal areas because it is detrimental to human health, economic development, the climate, and our planet's store of biodiversity.

The introduction of harmful pollutants into the marine environment interferes with the functioning of the marine ecosystem, and heavy metals are toxic to marine life such as fish and shellfish, and subsequently to humans. Nowadays, remote sensing technology has been developed and widely applied to assess marine pollution, and models have been developed to assess pollutants.

Remote sensing data combined with information from in-situ observations helps in the detection and extraction of polluted components in water, and accurate measurements of pollution levels in large regions ensures objectivity for methods of analysis.

This book discusses all methods, algorithms, and models that process and measure concentrations of pollutants. This publication will be valuable to marine biologists and environmentalists concerned with marine pollution and monitoring methods.

Houma Bachari Fouzia
National Higher School of Marine Sciences and Coastal Management,
Algeria

Chapter 1

Introductory Chapter: Marine Monitoring Pollution

Houma Fouzia, Boufeniza Redouane Larbi, Adem Amina, Chabi Nacera and Bachari Nour El Islam

1. Introduction

Monitoring the quality of the marine and coastal environment combines activities of various kinds and is defined as a type of activity that can be exercised on a regulatory basis (this is a control) or to evaluate levels or trends for a scientific study. This definition made it possible to clarify later, after a good number of debates, the definition of the monitoring objectives. It was at the origin of the extensive definition produced by the Oslo and Paris Conventions (the OSPAR Convention), which constitutes the most current reference: "continuous monitoring is the repeated measure of the quality of the marine environment and of each of its compartments, namely, water, sediment and living environment; natural or anthropogenic activities or inputs that may affect the quality of the marine environment; and the effects of its activities and contributions" [1].

Monitoring of the coastal and marine environment in particular requires the study of water (physical chemistry, temperature, salinity, oxygen, bacteriology, etc.), the sediment (grain size, micro, etc.), and living (benthos, plants, magnoliophytes, algae, fish, coral, biomonitoring, bioindicators). The methods and means of analysis and monitoring features of the marine and coastal environment (physical and chemical parameters, pollutants, nutrients, etc.) are numerous. Measurements are essential for understanding and interpreting data to accomplish the goals of surveillance [2].

The study of environmental pollution implies as a precise knowledge as possible of the distribution of pollutants in ecosystems and their effects on living organisms. Sometimes, it is customary to distinguish between a chemical monitoring whose purpose is to determine the level of contamination by a particular pollutant biotope and biomass and other biological monitoring which aims to assess the impact at a given moment or time of environmental pollution on exposed populations and communities. Since the critical level of ecotoxicological concentration-response relationship to a given pollutant is known, it will subsequently be possible to establish environmental protection standards for the pollutant under consideration.

2. Monitoring of general quality parameters

2.1 Enrichment and eutrophication parameters

This monitoring only covers water bodies. The basic parameters are temperature, salinity, nutrients (nitrate, nitrite, ammonium, phosphate), chlorophyll a, and pheopigments. On some sites, dissolved oxygen and silicate are also measured.

Analysis and determination of concentrations of pollutants environment can track their transfer to coastal waters.

The enrichment of water by nutrients, especially nitrogen and/or phosphorus, causes an accelerated growth of algae and higher forms of plant life to produce an undesirable disturbance to the balance of organisms present in the water and to the quality of water concerned. Thus, due to massive chemical inputs (agriculture, detergents, etc.), enrichment can lead to eutrophication. Therefore, it is important to follow the control parameters of this evolution (COD, BOD5, chlorophyll, organic matter, turbidity, etc.) [3].

Also, the optical properties (wavelengths of absorption and diffusion) of suspended particles are similar to those of chlorophyll pigments. The turbidity causes a loss of the luminous flux (by absorption or diffusion) in the water mass, which makes its quantification by modulation of its spectral signal a very important point for a better estimation of the Chl-a by satellite.

2.2 Chemical contaminants

2.2.1 Heavy metals

This contamination due to the presence of heavy metals and elements contaminating the biosphere, lead, as well as cadmium, arsenic and mercury, which is the most worrying of these pollutants. The contribution of the chemical species results in a more or less important contamination, that is to say a level of concentration above the normal, or a level at least detectable.

Contaminants are most often measured in living matter (shellfish and fish), in sediment, but sometimes also in water or suspended matter. The toxicity of a metal depends on its concentration in relation to the need and the tolerance of the organisms. At excessive concentrations, even trace elements can become toxic [4].

2.2.2 Hydrocarbons and PCB

The behavior of hydrocarbons at sea is complex: it obeys many physical and chemical processes. The oil is spreads on the surface of the sea (voluntarily or accidentally), and floats under the influence of its own weight, and the forces exerted on it are the current and the wind. These forces will also combine to move the tablecloth.

Over time, the lighter components of the oil will evaporate, and some will dissolve in the water. The waves, by their agitation, will cause a dispersion of tiny droplets in the water column or cause the formation of an emulsion of water and oil, which will stay on the surface.

The rate at which these different physicochemical phenomena occur is a function of the nature of the oil, wind speed, wave height, and temperature. When pollution occurs, it is very useful for surveillance teams to know how it will evolve. To do this, prediction models are used based on available information on pollution. These models also use weather and hydrodynamic forecasts. In marine ecosystems, PAHs and PCBs are bioavailable to fish and invertebrates (especially mollusks) not only from water and suspended matter in water but also from sediment [5].

3. Monitoring of biological effects

Monitoring biological effects aims to assess the health status of marine flora and fauna by measuring the response of these organisms to disturbances in the quality of the environment.

The microbiology of seawater concerns all microscopic living organisms (bacteria, viruses, fungi, etc.). The search for microorganisms can be done in water or shellfish as needed. Oysters and mussels are the most used ones (filtering organisms).

Fecal coliforms are used as indicators of fecal contamination of the water environment (freshwater and marine waters) and also indicative of a probable presence of pathogenic microorganisms. They are most often taken into account in surveillance. Bacteriological analysis is a means of determining the quality of seawater.

4. Environmental monitoring: satellite observation

Satellite remote sensing is considered a promising technique for studying some phytoplanktonic algae because of its advantages of large-scale, real-time, and long-term monitoring; the application of statistical models in the field of remote sensing is an indispensable tool. The main objective of this study was to quantify the spatial distribution, and develop empirical models, and detect phytoplankton algal bloom phenomena (diatoms and dinoflagellate), transparency of water, turbidity of sea, coastal pollution, chlorophyll, suspended matter, and hydrocarbons with remote sensor satellites.

Detection of changes in transparency of the water column is essential for understanding the responses of marine organisms to the availability of light. Water transparency models determine long-term geographic and depth distributions, while acute reductions cause short-term stress, potentially mortality, and may increase the vulnerability of organisms to other environmental factors. Several studies have shown that monitoring water clarity, which is well correlated with water quality, is a good way to manage water quality; however, it is impossible to monitor the clarity of water in a large area for long periods.

Also, thermal pollution of marine water adversely affects the local environment and ecosystems due to consequent rise in water temperature [5].

5. Aims of this book

Surveillance provides some of the scientific basis for standard development and application. The methodology of marine pollution control is governed by algorithms and models.

A monitoring strategy should be put in place, coupled with an environmental assessment concept, through targeted research activities in areas identified at the local and regional level. This concept will make it possible to diagnose the state of "health" of these zones and consequently to correct the anomalies.

6. The main objectives are:

6.1 Task I: ecological monitoring

Analysis and determination of environmental concentrations of pollutants and their transfer to coastal waters involves the developing ecological parameters to detect changes in ecosystem health. This includes new research on the field ecological status of species in addition to conventional monitoring and assessment of the ecological status of coastal waters.

6.2 Task II: risk assessment

The application of risk assessment techniques will make it possible to evaluate the impact of substances on the marine ecosystem.

Using bioaccumulation and biomagnification measurements, the potential effects of pollutants detected on humans and local populations of common terns will be assessed.

6.3 Task III: development of marine environmental management and monitoring tools

In this phase, integrated risk assessment procedures are developed for the assessment of the impact of pollutants on human health and the environment. Other tools for monitoring and mapping marine pollution are integrated into a marine information system that manages pollutant databases and simulations obtained by high-resolution temporal and spatial satellite imagery (maps, satellite images, pollutant concentration, hydrodynamic parameters, physicochemical parameters, phytoplankton, sea color, chlorophyll, temperature, turbidity, suspended matter, coastal bathymetry, salinity, etc.). The result of monitoring is a computerized decision support tool that meets the objectives of managers [6].

Author details

Houma Fouzia[1]*, Boufeniza Redouane Larbi[1], Adem Amina[1], Chabi Nacera[1] and Bachari Nour El Islam[2]

1 National Higher School of Marine Sciences and Coastal Management, Algeria

2 University of Science and Technology Houari Boumediene (USTHB), Algiers, Algeria

*Address all correspondence to: houmabachari@yahoo.fr

IntechOpen

© 2019 The Author(s). Licensee IntechOpen. This chapter is distributed under the terms of the Creative Commons Attribution License (http://creativecommons.org/licenses/by/3.0), which permits unrestricted use, distribution, and reproduction in any medium, provided the original work is properly cited. (cc) BY

References

[1] Fouzia H, Samir B, El Islam BN, Rabah B. Contribution of satellite measurements to the modeling and monitoring of the quality of coastal seawater. In: Perspectives in Water Pollution. Rijeka, Croatia: InTech Book, The Tamil University, Ocean and Atmospheric Sciences; 2013. 220 pages. Available from: http://www.intechopen.com/booksprocess/allchapters/chapter/84653

[2] Houma F, Belkessa R, Bachari NEI. Contribution of multispectral satellite imagery to the bathymetric analysis of sea bottom application to Algiers city, Algeria. Revue des Energies Renouvelables. 2006;**9**(3):165-172. Available from: http://www.cder.dz/?rub=revue&srub=n&pag=v9n1

[3] Bachouche S, Houma F, Gomiero A. Distribution and environmental risk assessment of heavy metal in surface sediments and red mullet (*Mullus barbatus*) from Algiers and BouIsmailBay (Algeria). Environmental Modeling and Assessment. 2017;**22**:1-18

[4] Bachari NEI, Houma F, Amarouche K. Combination of satellite images and numerical model to followed the coast of the bay of Bejaia-Jijel. International Journal of Environment and Geoinformatics. 2017;**4**:1-7

[5] Fouzia H, Ali K, Mostefa B. The development of a methodology to characterise and determine the sea water pollution by hydrocarbons using satellite images. In: The Mediterranean Science Commission, Rapp. Comm. int. Mer Médit. Espagne: CIESM; 2004. p. 37. Available from: http://www.ciesm.org/online/archives/abstracts/index.html

[6] El Islam BN, Fouzia H. Contribution of satellites visible and infrared images for the follow-up of the inshore water quality. In: International Conference on "Monitoring & Modeling of Marine Pollution" (INCOMP 2008); KISH. 2008. Available from: http://www.inoctr.org/proceedings---reports.html

Chapter 2

Detection and Monitoring of Marine Pollution Using Remote Sensing Technologies

Sidrah Hafeez, Man Sing Wong, Sawaid Abbas, Coco Y.T. Kwok, Janet Nichol, Kwon Ho Lee, Danling Tang and Lilian Pun

Abstract

Recently, the marine habitat has been under pollution threat, which impacts many human activities as well as human life. Increasing concerns about pollution levels in the oceans and coastal regions have led to multiple approaches for measuring and mitigating marine pollution, in order to achieve sustainable marine water quality. Satellite remote sensing, covering large and remote areas, is considered useful for detecting and monitoring marine pollution. Recent developments in sensor technologies have transformed remote sensing into an effective means of monitoring marine areas. Different remote sensing platforms and sensors have their own capabilities for mapping and monitoring water pollution of different types, characteristics, and concentrations. This chapter will discuss and elaborate the merits and limitations of these remote sensing techniques for mapping oil pollutants, suspended solid concentrations, algal blooms, and floating plastic waste in marine waters.

Keywords: remote sensing, water pollution detection and monitoring, optical sensors, oil spill, algal blooms, chlorophyll-a, suspended sediment concentration, marine plastic litter

1. Introduction

The oceans act as a natural sink for carbon dioxide and other greenhouse gases. However, anthropogenic activities have severely polluted the marine environment in the past few decades. Pollutants including plastic, oil, toxic chemicals, radioactive waste, and domestic and industrial sewage can be found in marine waters. Marine pollution is also caused by the discharge of sewage into rivers and excessive nutrients entering marine waters from agricultural fertilizers and pesticides [1]. These pollutants have adverse impacts on marine ecosystem including but not limited to sensitive coral reefs, mangroves, and aquaculture [2]. Therefore, in addition to reducing pollutant flow into oceans, it is essential to map and monitor marine pollutants to ensure a sustainable marine ecosystem.

Scientists and researchers have been working on detailed ocean monitoring for a sustainable blue economy. A variety of sensing systems are now available for ocean monitoring including autonomous underwater vehicles (AUVs), profiling floats, gliders, drifters, volunteer measurements from ships, and sensing nodes with cable networks [3]. These approaches to marine monitoring usually measure temperature, conductivity, pH, salinity, dissolved oxygen, fluorescence due to chlorophyll, turbidity, and color dissolved organic matter (CDOM). The most common approach for marine pollution measurements is to use conventional method of collecting in situ water samples using boats/ships from different depths of water with water samplers. The water samples are analyzed in the laboratory to determine the physical and chemical properties of the water. Such methods are accurate but time-consuming and geographically constrained and require trained professionals and laboratory analysis. However, real-time or near real-time measurements of marine pollutants and toxins across a range of spatial scales are necessary for monitoring and managing the environmental impacts and understanding the processes governing their spatial distribution [3].

To overcome these problems, remote sensing technology provides spatially synoptic and near real-time measurements that can be effectively used to detect, map, and track many pollutants such as oil and chemical spills, algal blooms, and high suspended solid concentrations. Aerial and satellite remote sensing has been demonstrated as an effective tool in detecting and mapping pollutant spills and for providing useful input data for oil spill models, to track pollutants through space and time [4–6]. An added advantage of remote sensing is that it provides information from remote areas. However, existing remote sensing technology still has some limitations, such as estimating pollutants over the vertical dimension of the water column.

The initial premise of watercolor remote sensing was to determine optical water quality variables such as chlorophyll-a (Chl-a) concentration, diffuse attenuation coefficient, and water-leaving radiance spectra [7]. The optical properties of water depend on many factors, e.g., suspended organic and inorganic particles and dissolved substances. There have been many successful applications of using remote sensing sensors for water color monitoring. The coastal zone color scanner (CZCS), having a spatial resolution of 825 m for six spectral bands from 443 to 750 nm, was the earliest satellite sensor designed and launched in 1978 specially to study ocean color. The sea-viewing wide field-of-view sensor (SeaWiFS) was the successor to CZCS with a spatial resolution of 1.1 km for eight spectral bands from 402 to 885 nm. Currently, many satellite sensors provide ocean color data for marine monitoring such as the moderate resolution imaging spectroradiometer (MODIS), the geostationary ocean color imager (GOCI), the visible infrared imager radiometer suite (VIIRS), the ocean and land color imager (OLCI), the Landsat operational land imager (OLI), and the Sentinel-2 multispectral instrument (MSI), all of which have suitable spectral and spatial resolutions capable of detecting marine pollutants and other water quality parameters (**Table 2**).

In order to track marine pollutants, prior understanding of marine dynamics is important, such as ocean current direction and magnitude, direction and speed of surface winds, sea surface temperature (SST), and sea surface salinity (SSS). Remote sensing now provides multiple satellite and airborne sensors to acquire information about marine dynamics over the vast marine regions. Apart from optical data, scanning radiometers and microwave sounders measure SST data, altimeters collect wave height data, and synthetic aperture radar (SAR) can measure the sea surface roughness patterns from which information on sea surface winds can be derived [31]. These datasets are of critical importance for detection and tracking of pollutants.

2. Remote sensing of water monitoring

Remote sensors capture the response of the electromagnetic interaction with water (**Figure 1**). Absorption and scattering are inherent optical properties (IOP) of water; and variations in IOP change the reflectance of water which is captured by a remote sensing sensor, and this is known as the apparent optical properties (AOP) of water (**Figure 2**). Reflection, absorption, and transmittance of electromagnetic radiation are highly dependent on the concentrations, types, and presence of substances in water. Total absorption is the sum of absorption by phytoplankton (microalgae), non-algal pigments (NAP), color dissolved organic matter (CDOM), and absorption by water, whereas light scattering by water is mainly controlled by suspended sediments (SS) present in water. Hence, ocean color represents the responses in , green, and red region, and data can be used to estimate the concentrations of water constituents [7].

Generally, clear water has low reflectance in the visible spectrum and has no reflection in near infrared (NIR) region, as it is absorbed by clear water. However, high reflectance measurements in red (600–700 nm) and NIR region (750–1400 nm) show a strong correlation with SS concentrations. A high concentration of suspended sediments blocks the transmittance to and from lower depths and therefore increases reflectance from the water surface. Similarly, high concentrations of chlorophyll (a photosynthetic pigment in phytoplankton and macroalgae) in water cause high reflectance in the green region (500–600 nm) and high absorption in the blue and red regions due to photosynthetic activity (**Figure 2**).

A portion of absorbed incident energy by the earth's features is also re-emitted in the thermal infrared region of the electromagnetic spectrum. Many satellite sensors such as MODIS, VIIRS, the advanced very high-resolution radiometer (AVHRR), and the sea and land surface temperature radiometer (SLSTR) measure the emitted thermal energy to determine sea surface temperature (SST). SST is an important parameter for understanding ocean water circulation. In case of large oil spills, these data can be effective for pinpointing the oil spilled areas, as they appear cooler than water surface due to their lower emissivity [31].

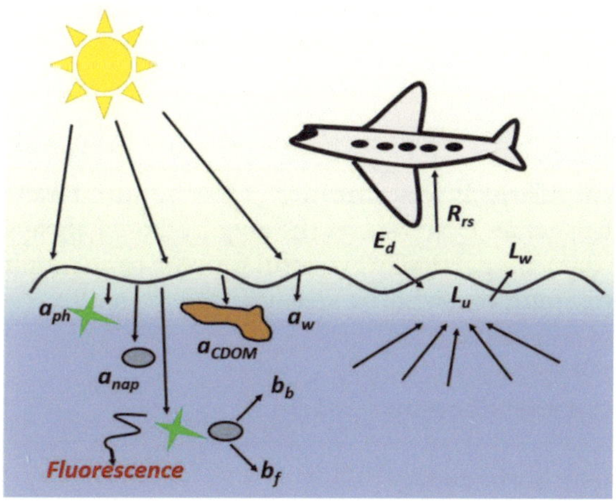

Figure 1.
Interaction of light with the water surface. a is absorption (a_{ph}, absorption by phytoplankton; a_{nap}, absorption by non-algal pigments; a_{CDOM}, absorption by color dissolved organic matter; and a_w, absorption by water), b is backscattering (b_b, backward scattering; b_f, forward scattering), R_{rs} is remote sensing reflectance recorded by sensor, E_d is downwelling irradiance, L_u is upwelling radiance, and L_w is water-leaving radiance [32].

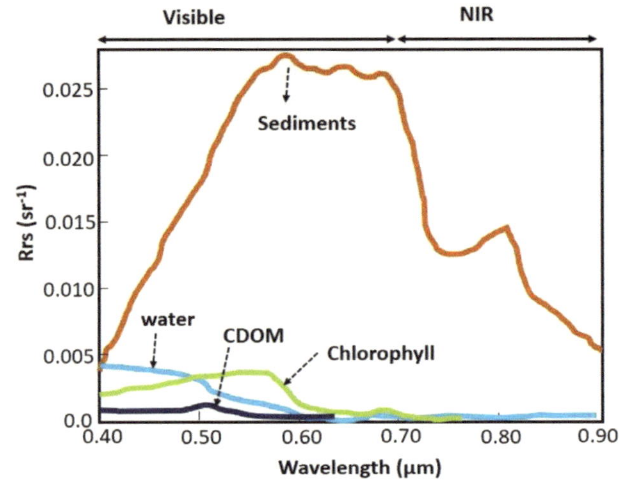

Figure 2.
Reflectance (R_{rs}) by clear water (blue), water with chlorophyll content (green), water with CDOM (black), and sedimented water (orange) [32].

Fluorescence is another type of energy emitted by a substance when it comes to a lower energy level from a higher energy level. The emitted energy is in a longer wavelength than the excitation wavelength. Algae absorb visible light for the photosynthesis process and emit excessive energy in the form of fluorescence signal (681 nm, the fluorescence band) when chlorophyll molecule comes to the non-excitation state during the photosynthesis process. The fluorescence can be detected by optical sensors with fine spectral resolution in the far-red and NIR and has a potential source for monitoring changes in the photosynthesis process in plants. Furthermore, in laser fluorometry, laser light is used to excite molecules [33]. This technique is common to detect oil and chemical spills [31].

2.1 Remote sensing platforms and sensors for water monitoring

There are now several remote sensing platforms for monitoring water pollutants, and they can be categorized into two types: airborne and spaceborne.

2.1.1 Airborne sensors

An aircraft flies at relatively low altitudes (a few hundred meters to a few kilometers above the surface); therefore, the acquired data always have higher levels of detail. Airborne data are particularly useful for real-time monitoring of oil and chemical spills. Four common airborne sensors used for spill surveillance [34] are listed below:

i. Infrared/ultraviolet line scan (IR/UVLS)

ii. Side-looking airborne radar (SLAR)

iii. Microwave radiometer (MWR)

iv. Laser fluorosensor (LF)

Airborne hyperspectral sensors with fine spatial resolution are able to capture detailed spectral variations. Therefore, they help to select the appropriate spectral region to study a specific water quality parameter, design satellite sensors, and improve already developed algorithms. Some airborne hyperspectral sensors particularly useful for studying coastal/river water quality are described in **Table 1**.

2.1.2 Spaceborne sensors

Spaceborne sensors can cover extensive and remote areas for water quality monitoring. Optical spaceborne sensors used for marine monitoring are mostly in sun-synchronous orbit; only GOCI, designed specifically for marine monitoring, is placed in geostationary orbit. The spatial coverage of these sensors ranges from tens to hundreds of kilometers, and the temporal frequency is from hourly to weekly monitoring.

Many algorithms have been developed to retrieve water quality information such as primary productivity, Chl-a variability, SS, total suspended solids (TSS), turbidity, total nitrogen, total phosphorus, CDOM, and SST. **Table 2** shows the satellite sensors most used for the study of water quality parameters related to marine pollution. The major application areas of active spaceborne sensors include, but are not limited to, sea surface currents, oil spills, biogenic films (algal blooms), and river plumes (**Table 5**).

Sensor	Manufacturer	Number of bands	Spectral range (nm)	Spatial resolution (m)	Studied parameter
Airborne visible infrared imaging spectrometer (AVIRIS)	NASA Jet Propulsion Lab	224	400–2500	17	Bottom albedo, water absorption, backscattering coefficients [35], Chl-a, CDOM, TSS [36]
HyMap	Earth Search Sciences Inc.	128	400–2500	3–10	Heavy metals [37]
Portable remote imaging spectrometer (PRISM)	NASA Jet Propulsion Lab	—	350–1050, SWIR band (1240 and 1640)	0.3	Sediment, CDOM, chlorophyll fluorescence [38] turbidity, Chl-a, dissolved organic carbon [39]
Airborne prism experiment (APEX)	VITO (Belgium)	313	VIS and NIR (380–970), SWIR (970–2500)	2–5	Chlorophyll fluorescence, SS [40]

Table 1.
Hyperspectral airborne sensors used in water quality assessment.

Satellite sensor	Launch date	Spectral bands (nm)	Spatial resolution (m)	Swath width (km)	Marine parameter accessed
Satellite sensors with moderate spatial resolution					
Landsats 4 and 5 TM	1 March 1984	5 (450–1750), 1 (2080–2350), and 1 (1040–1250)	30–120	185	Chl-a, SS, Secchi depth [8]
Landsat 7 ETM+	15 April 1999	6 (450–1750), 1 Pan (520–900), 1 (2090–2350), and 1 (1040–1250)	15–30–60	183	Chl-a, SS, Secchi depth, turbidity [9]
Terra Aster	18 December 1999	3 (520–860), 6 (1600–2430), and 5 TIR (8125–11,650)	15–30–90	60	Chl-a [10]
EO-1 ALI	November 2000	(443–2350)	30		Turbidity [11], SS [12]
EO-1 Hyperion	1 November 2000	242 (350–2570)	30	7.5	Chl-a, SS, CDOM [13, 14]
PROBA CHRIS	22 October 2001	19 (400–105)	18–36	14	Chl-a, phycocyanin [15] behenic macroalgae [16]
HICO	10 September 2009	128 (350–1080)	100	45–50	Chl-a, turbidity, CDOM [17], SS [18]
Landsat 8 OLI/ TIRS	11 February 2013	1 cirrus cloud detection (1360–1380), 5 (430–880), 1 Pan (500–680), 2 (1570–2290), 2 TIRS (10,600–12,510)	15–30–100	170	Chl-a, SS, turbidity, TN, TP [19]
Sentinel-2 MSI	23 June 2015	8 (490–865), 1(443) coastal aerosol, and 3 (1375–2190)	10–20–60	290	Chl-a, CDOM, DOC [20], SS [21]
Satellite sensors for regional coverage					
Orb View 2 SeaWiFS	1 August 1997	8 (402–885)	1130	2806	Chl-a [22]
Terra, Aqua MODIS	18 December 1999	2 (620–876), 5 (459–2155), 29 (405–877), and thermal	250–500–100	2330	Chl-a, CDOM SS [23], turbidity [24], TP [25]
ENVISAT-1 MERIS	1 March 2002	15 (390–1040)	300–1200	1150	Chl-a, SS [26, 27]
GOCI	26 June 2010	8 (412–865)	500	2500	Chl-a, SS, turbidity [28]
Suomi NPP VIIRS	28 October 2011	5 bands (640–1145), 16 bands (412–12,013), DNB (500–900)	375–750	3060	Chl-a [29]
Sentinel-3 OLCI	16 February 2016	21 (400–1020)	300	1270	Chl-a, SS, CDOM, and Secchi depth [30]

Table 2.
Satellite sensors mostly used to retrieve marine water quality parameters.

3. Remote sensing for marine monitoring

3.1 Chlorophyll (Chl-a) and algal blooms

Most algal species are nontoxic and are always present in coastal and open oceans. Planktons are the base of the marine food chain [22]. But, algae do not have to produce toxins to be harmful to the environment. The accelerated growth of algae produces a large amount of biomass which blocks sunlight and produces an anoxic or hypoxic condition (dissolved oxygen is depleted from the water column), which is hazardous to marine life. Algal blooms also affect coastal operations such as movement of ships, coastal tourism, and coastal sports (**Figure 3**). Algal blooms can persist from a few days to more than a month and spatially they may extend from a few meters to tens of kilometers.

The impact of algal blooms on marine life depends largely on the algal species involved. In situ field data collected using vessels are important for determining the algal species and level of toxicity during the bloom. However, field data are always limited for estimating the spatial extent as well as the dispersion. Detection of algal bloom by estimating the Chl-a concentrations using satellite imagery has been well-researched, as remote sensing has been used to observe ocean primary productivity since the launch of CZCS in 1978. High spatial and temporal resolutions are the main requirements of remote sensing data to study the variability in ocean and coastal Chl-a. By comparing a time series of satellite images, researchers can evaluate the spatial and temporal variations in Chl-a concentration during the bloom. This can also help to understand the dynamics of blooms. However, there are still certain conditions for using optical remote sensing to detect Chl-a, including (i) no or low cloud cover, (ii) the bloom should be near to the surface, and (iii) the bloom must cause the coloration of the water.

Optical remote sensing can observe the coloration of water due to algal pigments. In the open ocean, the color of water is mainly determined by phytoplankton; hence, it is relatively simple to develop algorithms using a bio-optical approach and remote sensing reflectance [22]. In the open ocean, Chl-a can be retrieved from the ratio of blue and green wavelengths as Chl-a absorption is sensitive to blue wavelength and reflectance peak occurs in the green wavelength region [22]. However, in coastal waters, the color of water also depends on organic matter such as NAP, CDOM, and inorganic solids, and consequently it is more complex to determine accurate Chl-a concentrations in coastal/turbid waters. Researchers have demonstrated that waters with increased Chl-a concentrations show a lower

Figure 3.
Spread of green algae along the coast of Qingdao in 2008, when summer Olympics was planned in this coast (source: Corey Sheran/Flickr) (right) and algae visible in MODIS false color image (shortwave, NIR, and Red) (source: MODIS rapid response project at NASA/GSFC) (left).

spectral response at short wavelengths especially in the blue wavelength regions [41]. This is due to increased absorption of red and blue wavelengths during photosynthetic process. **Figure 4** shows the reflectance of water with increasing Chl-a concentrations. Thus, in coastal waters, the red/NIR ratio is more effective for retrieval of Chl-a due to the presence of suspended solids and the increased spectral response of Chl-a pigments at longer wavelengths [43].

Narrow spectral bandwidth is a necessity for accurate retrieval of Chl-a concentrations [7]. The height of the spectral peak between 700 and 710 nm is used as a proxy for phytoplankton biomass [44]. Many researchers have used broad wavelength data (i.e., Landsat, HJ-1A/1B) as input to regression and neural network approaches for estimating Chl-a, achieving reasonable accuracy (70–90%) [9, 19, 45, 46]. **Table 3** shows some studies and datasets used to study Chl-a in marine regions. Lim and Choi [19] found that green and NIR bands of OLI are highly correlated with Chl-a (R = 0.71) in Korean waters. Nazeer and Nichol [46] also used the red/blue ratio to retrieve Chl-a with high accuracy (R = 0.85). Gurlin et al. [43] calibrated three models for Chl-a concentrations from 0 to 100 mg m^{-3} using two bands (red and NIR) of MERIS and MODIS reflectance data. They found that a simple two-band model achieved a higher accuracy than a complex three-band model. Moses et al. [51] also calibrated a red-NIR algorithm for high Chl-a concentrations in productive turbid waters. **Figure 5** shows Chl-a concentrations in highly turbid Pearl River Estuary and connecting rivers, derived using high-resolution MSI data with the method of Moses et al. [51].

Recently, machine learning approaches taking advantages of reflectance in all bands have also been applied using Landsat [45, 52] and GOCI data [28]. Our work also shows the potential use of Landsat TM, ETM+, and OLI with a machine learning approach to estimate Chl-a in coastal waters (**Figure 6**). We have evaluated three machine learning models to estimate Chl-a in the coastal waters of Hong Kong, of which artificial neural networks (ANN) performed best resulting in higher R (0.91) and lower RMSE (1.4 μg/L) than models based on support vector regression (SVR) and random forest (RF) algorithms. Chlorophyll indices such as the cyanobacteria index [53], maximum chlorophyll Index [54], and maximum peak height algorithm [55] have been demonstrated the robustness for detecting algal blooms and surface scum in coastal waters. Lunetta et al. [56] described the potential of using cyanobacteria index to measure cyanobacteria cell counts in bloom situations using MERIS data. Nazeer et al. [57] used board waveband band data (Landsat TM, ETM+, and HJ-1A/1B CCD) along with meteorological data as inputs to an artificial neural network model to map phytoplankton cell counts

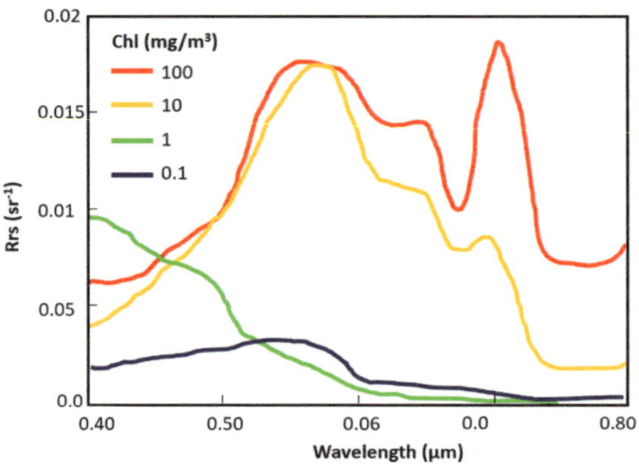

Figure 4.
Changing spectral response of water with different levels of chlorophyll concentration [42].

Band combination		**Sensor**	**Reference**
All bands (neural network and other machine learning methods)		GOCI	[28]
		TM, SAR	[45]
Multiple bands and their ratios (multiple regression)		OLI band (2–5)	[19]
		OLCI	[30]
		TM	[8]
		HICO	[17]
Blue (400–500 nm) and green (500–600 nm) ratio		In situ	[22]
Blue (400–500 nm) and red (600–700 nm) ratio		TM, ETM+, HJ-1A/1B CCD	[9, 46]
Green (500–600 nm) and red (600–700 nm) ratio		TM	[47]
		In situ (0.70/0.56 μm)	[44]
Red (600–700 nm) and NIR (700 μm–900 μm) ratio		MERIS, MODIS	[43]
		HICO	[48]
Using a single band	Green (500–600 nm)	Daedalus Airborne Thematic Mapper	[49]
	Red (600–700 nm)	AVHRR	[50]

Table 3.
Methods used to retrieve Chl-a using remote sensing data in the river and marine waters.

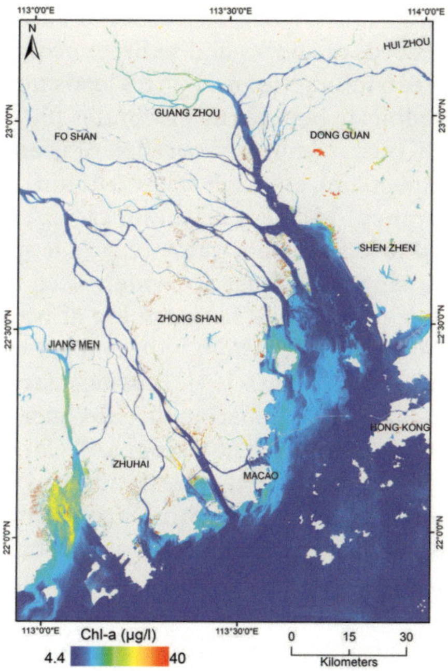

Figure 5.
Chl-a concentration observed in the Pearl River Estuary and its connecting rivers on 31 December 2017.

during a bloom in the complex coastal waters of Hong Kong and validated the model in two lakes in the United States and Japan.

Synthetic aperture radar (SAR) data can also be used to detect large algal blooms in cloudy weather as algal blooms may appear as an area of low backscatter compared to surrounding water surfaces [50].

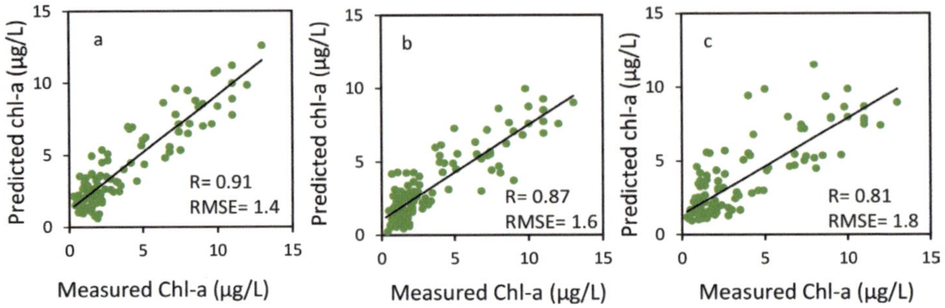

Figure 6.
Comparison of measured and predicted values from three machine learning models. (a) Chl-a concentration using artificial neural network, (b) Chl-a concentration using support vector regression, and (c) Chl-a concentration using random forest.

3.2 Turbidity, total suspended sediments, and stormwater runoff plumes

Turbidity is an optical property of water and is highly influenced by concentrations of suspended and dissolved organic and inorganic materials in water, including Chl-a, SS, and CDOM. SS is mainly responsible for the light scattering, whereas CDOM and Chl-a control the light absorption properties of water [58].

Turbidity and TSS are two important variables of marine systems studies because of their direct linkages with photosynthetically available radiation, which affects the growth of plankton and other algae [41]. Turbidity has also been used to measure fluvial SS concentrations in rivers and river plumes [59]. These fluvial SS loads are rich in nutrients and considered a cause of eutrophication. So, it is vital to have time series records of suspended sediment concentrations for better understanding of land-ocean interactions. High SS loads negatively affect aquaculture [59] and are hazardous to benthic invertebrates [60]. These parameters are also associated with the diffuse attenuation coefficient (penetration of light, in the blue-green region of the spectrum, through water column) and Secchi disk depth (a measure of water transparency) [41]. For all these reasons, turbidity and TSS concentrations are considered to be critical parameters in the study of marine systems.

Ocean color remote sensing techniques are widely used to monitor spatiotemporal variations in SS concentration and for mapping of water turbidity. **Figure 7** shows the changes in ocean color due to high sediment loads in the Yangtze River Estuary [60] and the Pearl River Estuary [61]. It is suggested that an algorithm using single bands provides a good estimation of TSS concentrations if an appropriate band is used [62]. Moreover Novo et al. [63] and Curran et al. [64] have demonstrated that a single-band approach may be adopted when water reflectance in the single band has a linear relationship with TSS concentrations. However, coastal water often consists of a complex mixture of substances and results in large variations in reflectance. In this case, multiple spectral bands should be adopted for TSS retrieval [62, 65, 66]. These methods using band arithmetic can achieve high accuracy around 80% for retrieving TSS concentrations in complex waters [67, 68]. The peak of the reflectance curve shifts from the green region to the red region with increasing concentration of dissolved and suspended matter; and water starts reflecting significantly in NIR region [21] (**Figure 8**). For water with high TSS concentrations, the spectral region between 600 and 900 nm should be used. Several studies using Landsat TM, ETM+, and OLI show that the blue, green, red, and NIR bands are useful for the determination of TSS [8, 19, 68–70]. Literature also shows that TM, ETM+, OLI, and MODIS are the most frequently used sensors for developing algorithms to study seasonal TSS variability in coastal and estuarine

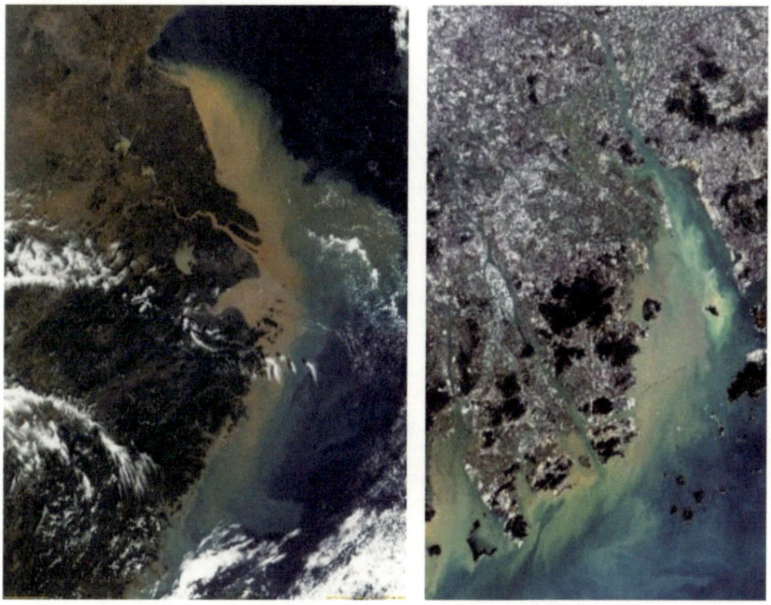

Figure 7.
Terra-MODIS true color image, captured on 16 September 2000, shows the sediment plume of the Yangtze River Estuary (left). The Sentinel-2 true color image, captured on 31 December 2017, shows high sediment concentrations in the Pearl River Estuary (right).

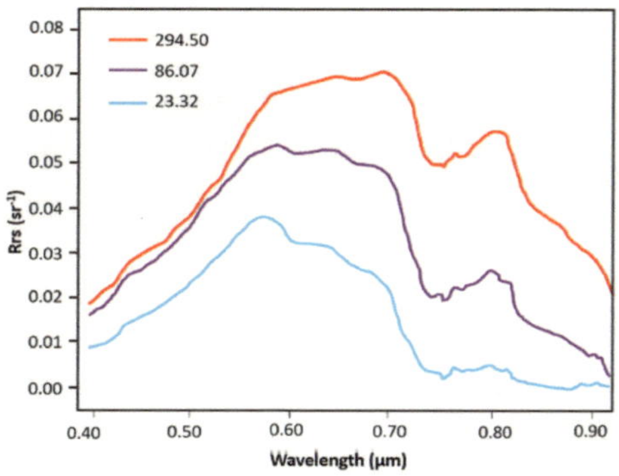

Figure 8.
Remote sensing reflectance (R_{rs}) spectra of water containing different suspended solid concentration (mg/L) [21].

areas, due to the large amount of archived remote sensing data [24, 71, 72]. The recently launched MSI sensor onboard Sentinel-2A and Sentinel-2B provide high spatial resolution of 10–20 m with a high temporal resolution of 5 days. The high spatial resolution (10 m) red and NIR bands are capable of routine monitoring of TSS concentration and turbidity in narrow bays, rivers, and inlets. **Figure 9** shows the suspended matter concentrations, and **Figure 10** shows turbidity in the Pearl River Estuary and connecting rivers using MSI data with algorithms of Nechad et al. [62] and Nechad et al. [73], respectively.

Methods and algorithms for estimation of TSS and turbidity have been evolved from simple methods such as linear/nonlinear regression and principal component analysis (PCA) to relatively complex techniques such as genetic algorithms and ANN. Nazeer and Nichol [68] initially developed a regression model resulting

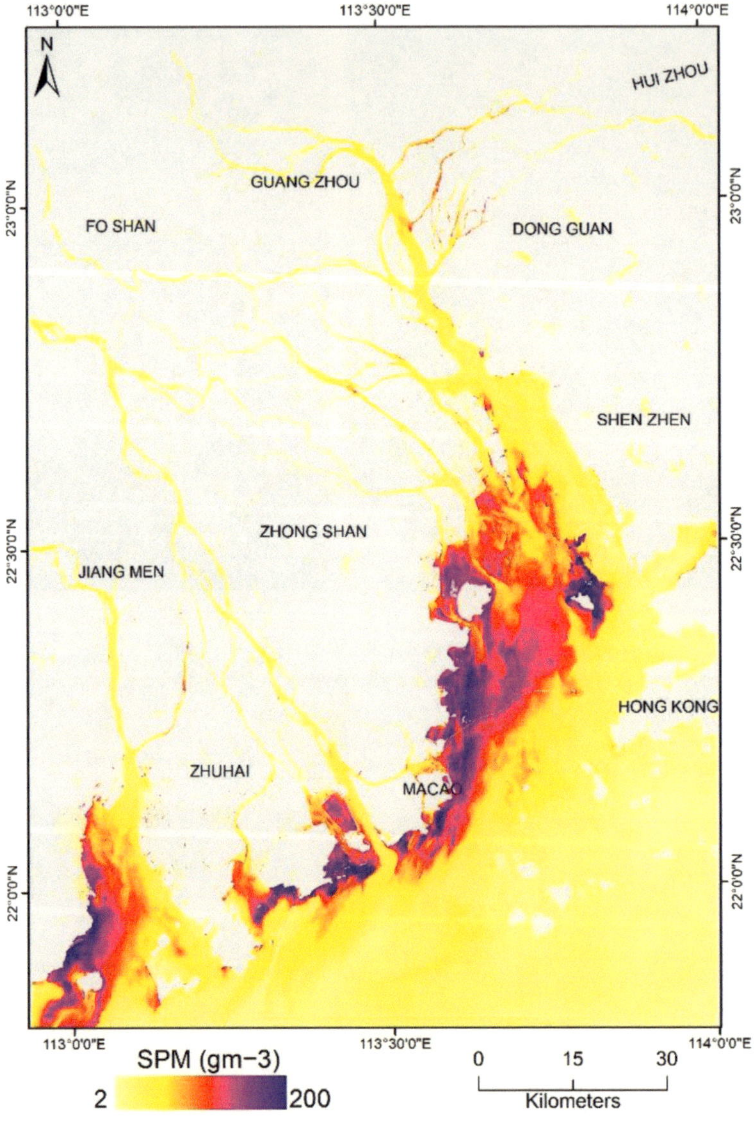

Figure 9.
High levels of suspended matter concentration were observed in the Pearl River Estuary and its connecting rivers on 31 December 2017.

in an RMSE of 2.60 mg/L. Later, Nazeer et al. [52] evaluated the potential of a machine learning model for estimating TSS in the complex coastal area of Hong Kong achieving an RMSE of 4.59 mg/L. Our work of machine learning models with Landsat TM, ETM+, and OLI data in the same area also shows promising results for estimation of TSS (**Figure 11**). In our work, ANN outperformed the other two machine learning approaches, SVR (support vector machine) and RF (random forest), resulting in the lowest RMSE of 2.8 mg/L. **Table 4** includes some studies and methods used to study TSS in rivers, bays, estuaries, and relatively open coastal waters.

Stormwater runoff is also a large source of marine pollution as runoffs and pollutants from the urban watershed enter into the coastal environment after rainstorms. Stormwater runoff and municipal wastewater plumes may sometimes be overlooked due to persistent cloud cover in optical imagery. These types of

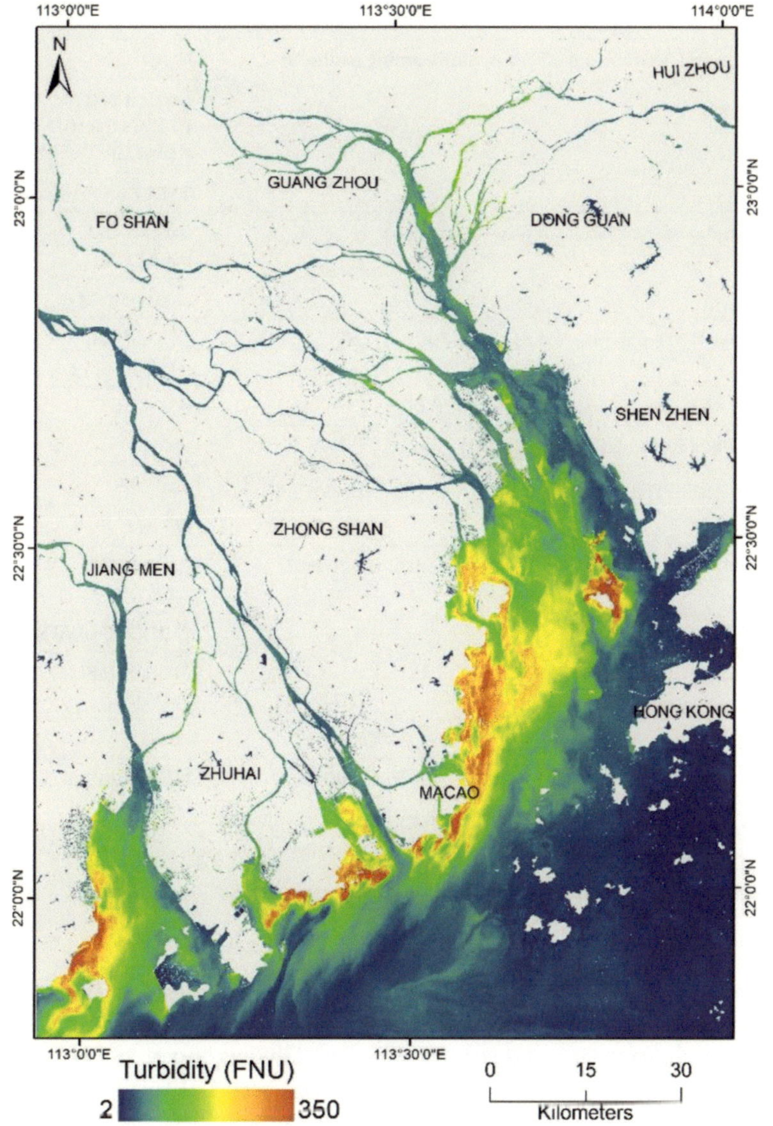

Figure 10.
High levels of turbidity were observed in the Pearl River Estuary and its connecting rivers on 31 December 2017.

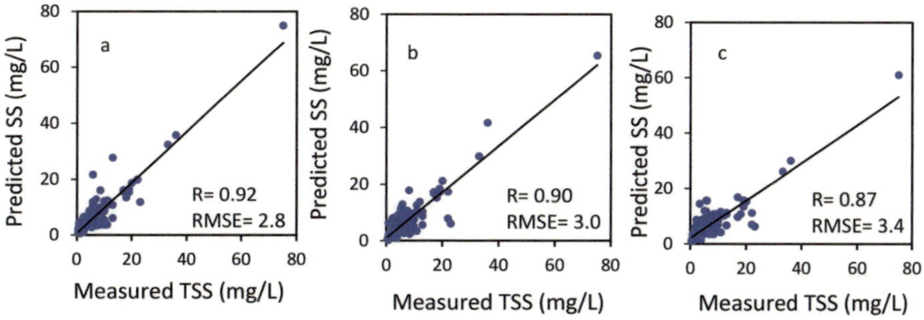

Figure 11.
Comparison of measured and predicted values from three machine learning models. (a) TSS concentration using artificial neural network, (b) TSS concentration using support vector regression, and (c) TSS concentration using random forest.

Band combination		Sensor	Reference
All bands (neural network and other machine learning methods)		GOCI	[28]
		Landsat TM, ETM+, OLI, HJ-1 A/B CCD	[52]
		TM, SAR	[45]
Multiple bands and their ratios (multiple regression)		Landsat OLI band (2–5)	[19]
		Landsat ETM+	[9]
Multiple green (500–600 nm) and red (600–700 nm) ratio		Landsat TM, ETM+	[68]
Green (500–600 nm) and red (600–700 nm) ratio		HJ-1A/1B CCD	[67]
Red (600–700 nm) and NIR (700–900 nm) ratio		MODIS	[65]
Single band algorithms	Green (500–600 nm)	SeaWiFS	[58]
		EO-ALI	[12]
	Red (600–700 nm)	Landsat TM, ETM+, HJ-1	[47, 68]
		AVHRR, SeaWiFS	[58]
		MODIS, MERIS, SeaWiFS	[24, 62, 65]
		HICO	[17]
	NIR (700–900 nm)	MODIS, MERIS, SeaWiFS	[62]

Table 4.
Methods used to retrieve TSS using remote sensing data in marine waters.

runoff are often detectable via SAR as they deposit surfactants on the sea surface, smoothing the small gravity waves and thus producing an area of low backscatter in comparison to the surrounding sea surface [74]. DiGiacomo et al. [74] used high-resolution SAR to monitor such plumes in the Southern California Bight. In their study, the dynamics of runoff plume was modeled using SAR images together with meteorological data as a function of cumulative event discharge, timing of the peak flow, and total storm precipitation. Holt et al. [75] used multi-platform SAR data along with MODIS and precipitation data to study a stormwater plume and its flow direction.

3.3 Oil spill

A large oil spill from tankers causes not only significant economic loss but also destruction to the aquatic ecosystem. After the spill, oil undergoes several processes such as spreading, evaporation, dissolution, drifting, photolysis, biodegradation, and the formation of oil-in-water and water-in-oil emulsions [76].

Owing to the dynamic spreading nature of the spill, both remote and station-based sensors are essential for comprehensive and effective monitoring. Airborne survey of an oil spill can be carried out by side-looking airborne radar (SLAR), laser fluorosensor (LF), and ultraviolet and thermal infrared video cameras. Ultraviolet, microwave, thermal, and optical airborne sensors all exhibit the ability to detect oil spills [6]. Ultraviolet sensors are sensitive to oil thickness of 0.01–0.05 μm. Oil appears as a bright target in this region of the spectrum, and brightness increases with the thickness of the oil. Optical sensors can measure thicker oil (2–500 μm) and are able to detect oil dispersed in water, whereas thermal infrared sensors measure oil with a thickness of about 10–50 μm [34]. Airborne LF and microwave radiometers (MWR) are considered to be the most appropriate sensors for oil spill detection. SLAR, ultraviolet, and thermal video cameras were used to identify areas of thick oil during the Sea Empress oil spill in 1996. Oil also undergoes weathering and aging. Multispectral satellite images, taking advantage of fluorescence characteristics of oil, can detect spills and assess the levels of weathering of the oil [31].

Spaceborne synthetic aperture radar (SAR) is commonly used for ocean pollution monitoring, especially oil spills. **Table 5** includes some SAR-equipped satellites used for oil spill detection. The advantage of SAR is the capability to take measurements during all day and all-weather conditions. Therefore, they are considered superior to optical sensors in this application [5]. The spreading trend of oil highly depends on wind direction and speed. An oil spill would break up and disperse if the wind speed is greater than 10 m/s [74]. DiGiacomo et al. [74] used ERS-2 SAR and RADARSAT-1 SAR images to map oil spills in the Southern California Bight. Shirvany et al. [77] evaluated the potential of different polarizations using RADARSAT-2 data for oil spill detection in the Gulf of Mexico. In another study, ENVISAT data was used effectively as an input to a hydrodynamic model to track the fate of oil after the Kerch Strait oil spill in 2007 [78]. **Figure 12** shows an incident of large oil spill on the Galicia coast [79] and the Korean coast [80] for which spaceborne SAR data was used to access the coverage areas and the damage caused by the spills.

3.4 Marine plastic and coastal litter

With the increasing amount of marine plastic litter, its adverse chemical, biological, and ecological impacts on the marine ecosystem have raised the public concerns [81]. It is estimated that 4.8–12.7 million metric tons of plastic is dumped in the sea every year [82] due to increased use of plastic in industry and daily life [83, 84]. Although some surveys have been undertaken [85] to estimate the density and weight of floating plastic in the oceans globally, there is a lack of long-term and large-scale monitoring.

Some research has been conducted using remote sensing technology for the detection of floating marine plastic [86]. However, this research domain is still in its early stages. The reflectance from water captured by sensors is different from that of floating plastic objects. There are several reasons for this, (1) the physical properties of water are different from that of plastic, and they have significant distinct reflectance; (2) the transmitting ability of light through water is different

Satellite sensor	Operation
Sentinel-1A	2014, operating
Sentinel-1B	2016, operating
TerraSAR-X	2007, operating
ENVISAT advanced synthetic aperture radar (ASAR)	2002, not operating
RADARSAT-1	1995, not operating
European remote sensing (ERS) satellites: ESR-2	1995, not operating

Table 5.
Active spaceborne sensors mostly used in oil spill detection.

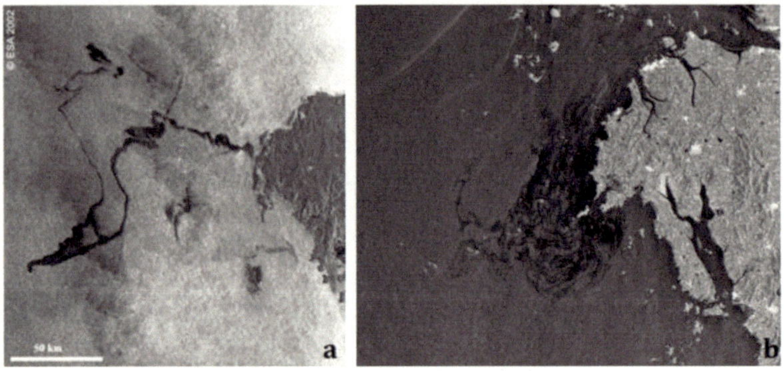

Figure 12.
(a) ASAR wide-swath image of northwest coast of Spain, captured on 17 Nov 2002, at 10:45 UTC showing oil from the wrecked tanker approaching Spanish coast (source, ESA), (b) ASAR image of South Korea, captured on 11 Dec 2007, at 01:40 UTC, showing oil spill from 146,000 ton damaged crude oil tanker (source ESA).

from that through plastic; (3) the absorption of light by water is different from plastic [87]. **Figure 13** shows different pathways of incident light after interacting with the surface (with and without marine plastic). Some studies have used hyperspectral remote sensing to study marine macroplastics [87] and microplastics [88]. Goddijn-Murphy et al. [87] considered the spectral signatures and geometric optics of plastic and seawater to develop a reflectance model for detecting macroplastics. The key is to determine the appropriate reflectance peak of plastic and consider its ratio with wavelength bands where water-leaving reflectance is low. Their model considers reflectivity of only one type of plastic litter in two dimensions. However, there are some constraints for detecting marine plastics in a real scenario since there have no standard shape, dimension, color, chemical composition, etc. Nevertheless, this study demonstrated the possibility of using remote sensing as a useful means for mapping and tracking of marine plastic.

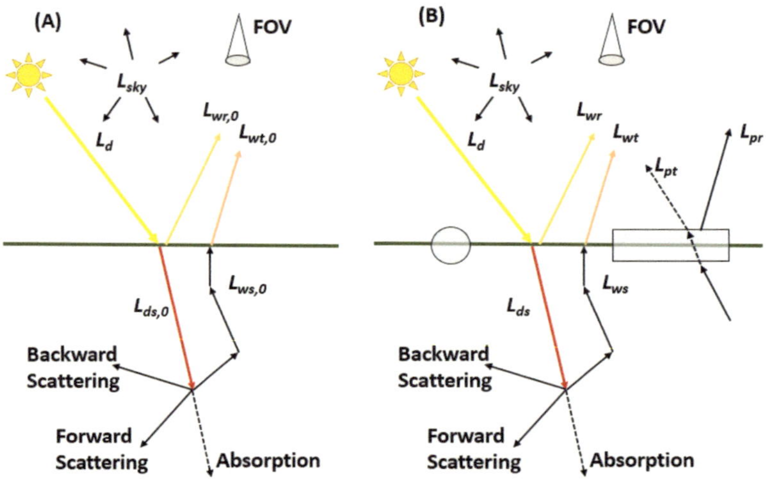

Figure 13.
Schematic of solar radiance interacting with (A) an open water body and (B) the same water body but with floating plastic. L_d is total downwelling radiance (solar beam + diffuse skylight), L_{ds} is subsurface downwelling radiance, L_{ws} is subsurface upwelling radiance, L_{wr} is radiance reflected directly off the water surface, L_{wt} is subsurface upwelling radiance transmitted through the water-air interface, L_{pr} is radiance reflected off the plastic, and L_{pt} is subsurface upwelling radiance transmitted through the plastic. L_w is total water-leaving radiance, L_{wr} + L_{wt}, and L_p is total plastic leaving radiance, L_{pt} + L_{pr}; subscript 'o' indicates all the variables in the absence of plastic and FOV is a field of view [87].

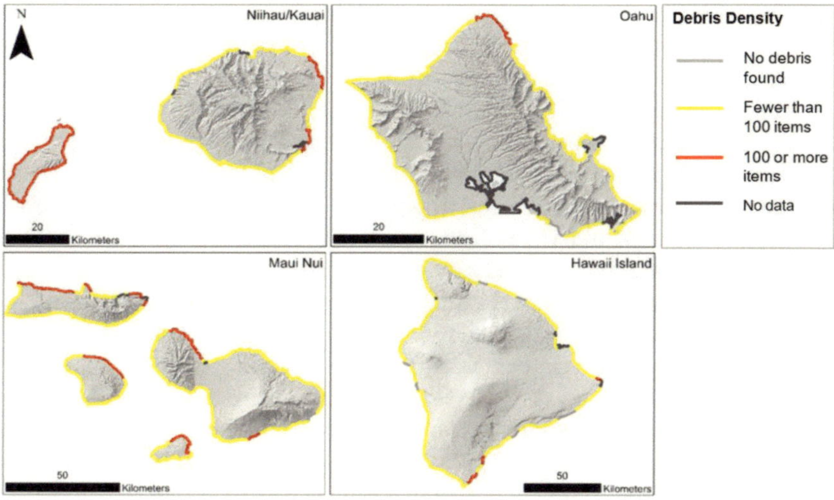

Figure 14.
Distribution and density of marine litter along the coasts of the main Hawaiian Islands. Areas with 100 and more item densities are shown as hotspots of high marine litter [89].

Detecting coastal litter near land surface is easier than in open ocean, as its reflectance and shape characteristics are not affected by its pitching and rolling on ocean waves. Moy et al. [89] used aerial imagery along with spatial analysis to categorize and map marine litter deposited along the coasts of the Hawaiian Islands. Very high-resolution aerial imagery allowed precise measurements of the quantity, location, type, and size of dumped litter (>$0.05\ m^2$) (**Figure 14**). In another study, Martin et al. [90] discussed the potential of combining images from unmanned aerial vehicles (UAV) and a machine learning approach, to detect and map marine litter. Machine learning algorithms are able to detect and classify objects when training samples with known training objects are provided. Their results showed that a UAV-based beach survey is 39 times faster than beach screening on foot and the large footprint of a UAV can cover entire coastlines and beaches including those in remote areas.

4. Conclusion

Increased levels of marine pollution due to anthropogenic activities are adversely affecting marine sustainability of marine ecosystems. Reviewed literature suggested that aerial and spaceborne sensors provide holistic information for monitoring many of the major marine pollutants. These include oil and chemical spills, sewage, high suspended solids, and algal blooms. Solid waste deposited in coastal areas can also be mapped using similar geospatial technology. However, there are some technical limitations in assessing detailed information about pollutants. These limitations stem from their dynamic nature, limited information of specific spectral response of pollutants, substrate response in optically shallow waters, and complex physics of light interaction through the water column. Despite these limitations, remote sensing is still capable of providing useful information about pollution events in sensitive marine areas.

Active and hyperspectral airborne sensors are often considered superior to spaceborne sensors for monitoring coastal and estuarine pollutants due to their real-time and detailed monitoring capability. Spaceborne sensors are more reliable for large-scale ocean, but with the recent development of sensor technology, especially hyperspectral and active sensors with high temporal resolution, the applications of spaceborne sensors in coastal regions are also increasing. Presently, monitoring of marine waters is offered through numerous satellite sensors such as MODIS, VIIRS, AVHRR, OLCI, GOCI, Landsat, and Sentinel-2 with spectral and spatial resolutions able to measure marine pollutants and other marine parameters. Active satellite sensors such as SAR, altimeters, scanning radiometers, and microwave sounders, which are mostly used in physical oceanography, also possess the potential for detection of marine pollution under specific meteorological conditions and provide useful data to track and model the impact of these pollutants.

Heavy metal pollution in coastal and estuarine region is another major concern of marine managers and researchers. Studies have attempted to use airborne hyperspectral data for this task, but satellite remote sensing is not yet able to detect these loads directly. However, the core factors causing these pollutants such as river plumes, sewerage, and industrial waste entering into these sensitive systems can be monitored using satellite remote sensing. If the point source of heavy metals is traced by remote sensing, policies and management practices can be applied according to the specific pollutants, and their mobilization and transfer of heavy metal to sensitive coastal environments can be avoided. Multiple approaches have proven reliable for this task.

In addition, recent developments in software and computation power have led to the increased use of data captured by remote sensing systems. Computer systems can now store and analyze large datasets. Therefore, marine protection agencies and government can utilize the full potential of remote sensing data in geographic information systems (GIS) and decision support systems (DSS) to manage marine resources and pollution. Collaboration between the research community and government is of utmost importance for using the full potential of this data in marine pollution management. Different applications of remote sensing such as detection of floating marine plastic litter and the use of active remote sensing for detecting algal blooms are still in the research. With the advancement of remote sensing sensors, sophisticated methods will be developed in the future for monitoring marine pollution.

Acknowledgements

Authors would like to acknowledge the General Research Fund (project id: 15246916), the Hong Kong Ph.D. Fellowship Scheme from the Research Grants Council of Hong Kong. The authors would also like to acknowledge US Geological Survey for providing Landsat (TM, ETM+, and OLI) image archive, the Copernicus Open Access Hub for providing Sentinel-2 data, and the Hong Kong Environmental Protection Department for providing station-based coastal water quality data for developing numerical models.

Author details

Sidrah Hafeez[1], Man Sing Wong[1*], Sawaid Abbas[1], Coco Y.T. Kwok[1], Janet Nichol[1], Kwon Ho Lee[2], Danling Tang[3] and Lilian Pun[1]

1 Department of Land Surveying and Geo-Informatics, The Hong Kong Polytechnic University, Kowloon, Hong Kong

2 Department of Atmospheric and Environmental Sciences, Gangneung-Wonju National University Gangneung, South Korea

3 South China Institute of Oceanology Chinses Academy of Sciences, China

*Address all correspondence to: lswong@polyu.edu.hk

IntechOpen

© 2018 The Author(s). Licensee IntechOpen. This chapter is distributed under the terms of the Creative Commons Attribution License (http://creativecommons.org/licenses/by/3.0), which permits unrestricted use, distribution, and reproduction in any medium, provided the original work is properly cited.

References

[1] Clark RB, Frid C, Attrill M. Marine Pollution. Vol. 4. Oxford: Clarendon Press; 1989

[2] Islam MS, Tanaka M. Impacts of pollution on coastal and marine ecosystems including coastal and marine fisheries and approach for management: A review and synthesis. Marine Pollution Bulletin. 2004;**48**:624-649

[3] Zielinski O, Busch JA, Cembella AD, Daly KL, Engelbrektsson J, Hannides AK, et al. Detecting marine hazardous substances and organisms: Sensors for pollutants, toxins, and pathogens. Ocean Science. 2009;**5**:329-349

[4] ESA. Sentinel-1 Supports Detection of Illegal Oil Spills. 2017. Available from: https://sentinel.esa.int/web/sentinel/missions/sentinel-1/news/-/article/sentinel-1-supports-detection-of-illegal-oil-spills

[5] Brekke C, Solberg AH. Oil spill detection by satellite remote sensing. Remote Sensing of Environment. 2005;**95**:1-13

[6] Zielinski O, Hengstermann T, Robbe N. Detection of oil spills by airborne sensors. In: Marine Surface Films. Berlin, Heidelberg: Springer; 2006. pp. 255-271

[7] Dekker AG, Brando VE, Anstee JM, Pinnel N, Kutser T, Hoogenboom EJ, et al. Imaging spectrometry of water. In: Imaging Spectrometry. Dordrecht: Springer; 2002. pp. 307-359

[8] Pattiaratchi C, Lavery P, Wyllie A, Hick P. Estimates of water quality in coastal waters using multi-date Landsat Thematic Mapper data. International Journal of Remote Sensing. 1994;**15**:1571-1584

[9] Kabbara N, Benkhelil J, Awad M, Barale V. Monitoring water quality in the coastal area of Tripoli (Lebanon) using high-resolution satellite data. ISPRS Journal of Photogrammetry and Remote Sensing. 2008;**63**:488-495

[10] Nas B, Karabork H, Ekercin S, Berktay A. Mapping chlorophyll-a through in-situ measurements and Terra ASTER satellite data. Environmental Monitoring and Assessment. 2009;**157**:375-382

[11] Chen S, Fang L, Zhang L, Huang W. Remote sensing of turbidity in seawater intrusion reaches of Pearl River Estuary—A case study in Modaomen water way, China. Estuarine, Coastal and Shelf Science. 2009;**82**:119-127

[12] Fang L, Chen S, Wang H, Qian J, Zhang L. Detecting marine intrusion into rivers using EO-1 ALI satellite imagery: Modaomen Waterway, Pearl River Estuary, China. International Journal of Remote Sensing. 2010;**31**:4125-4146

[13] Brando VE, Dekker AG. Satellite hyperspectral remote sensing for estimating estuarine and coastal water quality. IEEE Transactions on Geoscience and Remote Sensing. 2003;**41**:1378-1387

[14] Zhu W, Yu Q. Inversion of chromophoric dissolved organic matter from EO-1 hyperion imagery for turbid estuarine and coastal waters. IEEE Transactions on Geoscience and Remote Sensing. 2013;**51**:3286-3298

[15] Ruiz-Verdú A, Domínguez-Gómez J-A, Peña-Martínez R. Use of CHRIS for monitoring water quality in Rosarito reservoir. In: Proceedings of the Third Chris Proba Workshop. ESA-ESRIN; 2005

[16] Casal G, Kutser T, Domínguez-Gómez J, Sánchez-Carnero N, Freire J. Mapping benthic macroalgal communities in the coastal zone

using CHRIS-PROBA mode 2 images. Estuarine, Coastal and Shelf Science. 2011;**94**:281-290

[17] Keith DJ, Schaeffer BA, Lunetta RS, Gould RW Jr, Rocha K, Cobb DJ. Remote sensing of selected water-quality indicators with the hyperspectral imager for the coastal ocean (HICO) sensor. International Journal of Remote Sensing. 2014;**35**:2927-2962

[18] Braga F, Giardino C, Bassani C, Matta E, Candiani G, Strömbeck N, et al. Assessing water quality in the northern Adriatic Sea from HICO™ data. Remote Sensing Letters. 2013;**4**:1028-1037

[19] Lim J, Choi M. Assessment of water quality based on Landsat 8 operational land imager associated with human activities in Korea. Environmental Monitoring and Assessment. 2015;**187**:384

[20] Toming K, Kutser T, Laas A, Sepp M, Paavel B, Nõges T. First experiences in mapping lake water quality parameters with Sentinel-2 MSI imagery. Remote Sensing. 2016;**8**:640

[21] Liu H, Li Q, Shi T, Hu S, Wu G, Zhou Q. Application of sentinel 2 MSI images to retrieve suspended particulate matter concentrations in Poyang Lake. Remote Sensing. 2017;**9**:761

[22] O'Reilly JE, Maritorena S, O'brien MC, Siegel DA, Toole D, Menzies D. Chavez FP. In: SeaWiFS postlaunch calibration and validation analyses. part 3. NASA tech. memo. 2000;**206892**(11):3-8

[23] Miller RL, McKee BA. Using MODIS Terra 250 m imagery to map concentrations of total suspended matter in coastal waters. Remote Sensing of Environment. 2004;**93**:259-266

[24] Chen Z, Hu C, Muller-Karger F. Monitoring turbidity in Tampa Bay using MODIS/Aqua 250 m imagery. Remote Sensing of Environment. 2007;**109**:207-220

[25] Chang N-B, Xuan Z, Yang YJ. Exploring spatiotemporal patterns of phosphorus concentrations in a coastal bay with MODIS images and machine learning models. Remote Sensing of Environment. 2013;**134**:100-110

[26] Shen F, Verhoef W, Zhou Y, Salama MS, Liu X. Satellite estimates of wide-range suspended sediment concentrations in Changjiang (Yangtze) estuary using MERIS data. Estuaries and Coasts. 2010;**33**:1420-1429

[27] Harvey ET, Kratzer S, Philipson P. Satellite-based water quality monitoring for improved spatial and temporal retrieval of chlorophyll-a in coastal waters. Remote Sensing of Environment. 2015;**158**:417-430

[28] Kim YH, Im J, Ha HK, Choi J-K, Ha S. Machine learning approaches to coastal water quality monitoring using GOCI satellite data. GIScience & Remote Sensing. 2014;**51**:158-174

[29] Wang M, Son S. VIIRS-derived chlorophyll-a using the ocean color index method. Remote Sensing of Environment. 2016;**182**:141-149

[30] Toming K, Kutser T, Uiboupin R, Arikas A, Vahter K, Paavel B. Mapping water quality parameters with sentinel-3 ocean and land colour instrument imagery in the Baltic Sea. Remote Sensing. 2017;**9**:1070

[31] Loughland RA, Saji B. Remote sensing: A tool for managing marine pollution in the Gulf. In: Protecting the Gulf's Marine Ecosystems from Pollution. Birkhäuser Basel: Springer; 2008. pp. 131-145

[32] Sherry PL. Applied Remote Sensing Training [Internet]. Available from: https://arset.gsfc.nasa.gov/sites/default/files/users/fundamentals/fundamentals-aquatic-web.pdf

[33] Kaye TG, Falk AR, Pittman M, Sereno PC, Martin LD, Burnham DA, et al. Laser-stimulated fluorescence in paleontology. PLoS One. 2015;**10**:e0125923

[34] Theo Hengstermann NR. Airborne Oil Spill Remote Sensing. Hydro International. 2008. Available from: https://www.hydro-international.com/content/article/airborne-oil-spill-remote-sensing

[35] Lee Z, Carder KL, Chen RF, Peacock TG. Properties of the water column and bottom derived from Airborne Visible Infrared Imaging Spectrometer (AVIRIS) data. Journal of Geophysical Research, Oceans. 2001;**106**:11639-11651

[36] Lunetta RS, Knight JF, Paerl HW, Streicher JJ, Peierls BL, Gallo T, et al. Measurement of water colour using AVIRIS imagery to assess the potential for an operational monitoring capability in the Pamlico Sound Estuary, USA. International Journal of Remote Sensing. 2009;**30**:3291-3314

[37] Choe E, van der Meer F, van Ruitenbeek F, van der Werff H, de Smeth B, Kim K-W. Mapping of heavy metal pollution in stream sediments using combined geochemistry, field spectroscopy, and hyperspectral remote sensing: A case study of the Rodalquilar mining area, SE Spain. Remote Sensing of Environment. 2008;**112**:3222-3233

[38] Mouroulis P, Van Gorp B, Green RO, Dierssen H, Wilson DW, Eastwood M, et al. Portable remote imaging spectrometer coastal ocean sensor: Design, characteristics, and first flight results. Applied Optics. 2014;**53**:1363-1380

[39] NASA. NASA Demonstrates Airborne Water Quality Sensor. 2016. Available from: https://climate.nasa.gov/news/2404/nasa-demonstrates-airborne-water-quality-sensor/

[40] A.-A. P. EXperiment. APEX—Airborne Prism EXperiment Flyer. ESA. Winterthurerstrasse, Zurich, Switzerland: University of Zurich; 2012. pp. 5-8

[41] Gholizadeh MH, Melesse AM, Reddi L. A comprehensive review on water quality parameters estimation using remote sensing techniques. Sensors. 2016;**16**:1298

[42] Moore T. Challenges for Bio-Optical Modeling of Inland Waters. 2017. Available from: https://iocs.ioccg.org/wp-content/uploads/2017/05/tue-1445-bo4-moore.pdf

[43] Gurlin D, Gitelson AA, Moses WJ. Remote estimation of chl-a concentration in turbid productive waters—Return to a simple two-band NIR-red model? Remote Sensing of Environment. 2011;**115**:3479-3490

[44] Gitelson A. The peak near 700 nm on radiance spectra of algae and water: Relationships of its magnitude and position with chlorophyll concentration. International Journal of Remote Sensing. 1992;**13**:3367-3373

[45] Zhang Y, Pulliainen J, Koponen S, Hallikainen M. Application of an empirical neural network to surface water quality estimation in the Gulf of Finland using combined optical data and microwave data. Remote Sensing of Environment. 2002;**81**:327-336

[46] Nazeer M, Nichol JE. Development and application of a remote sensing-based Chlorophyll-a concentration prediction model for complex coastal waters of Hong Kong. Journal of Hydrology. 2016;**532**:80-89

[47] Hellweger F, Schlosser P, Lall U, Weissel J. Use of satellite imagery for water quality studies in New York Harbor. Estuarine, Coastal and Shelf Science. 2004;**61**:437-448

[48] Moses WJ, Bowles JH, Corson MR. Expected improvements in the quantitative remote sensing of optically complex waters with the use of an optically fast hyperspectral spectrometer—A modeling study. Sensors. 2015;**15**:6152-6173

[49] George D. The airborne remote sensing of phytoplankton chlorophyll in the lakes and tarns of the English Lake District. International Journal of Remote Sensing. 1997;**18**:1961-1975

[50] Svejkovsky J, Shandley J. Detection of offshore plankton blooms with AVHRR and SAR imagery. International Journal of Remote Sensing. 2001;**22**:471-485

[51] Moses WJ, Gitelson AA, Berdnikov S, Saprygin V, Povazhnyi V. Operational MERIS-based NIR-red algorithms for estimating chlorophyll-a concentrations in coastal waters—The Azov Sea case study. Remote Sensing of Environment. 2012;**121**:118-124

[52] Nazeer M, Bilal M, Alsahli MM, Shahzad MI, Waqas A. Evaluation of empirical and machine learning algorithms for estimation of coastal water quality parameters. ISPRS International Journal of Geo-Information. 2017;**6**:360

[53] Wynne T, Stumpf R, Tomlinson M, Warner R, Tester P, Dyble J, et al. Relating spectral shape to cyanobacterial blooms in the Laurentian Great Lakes. International Journal of Remote Sensing. 2008;**29**:3665-3672

[54] Gower J, King S, Borstad G, Brown L. Detection of intense plankton blooms using the 709 nm band of the MERIS imaging spectrometer. International Journal of Remote Sensing. 2005;**26**:2005-2012

[55] Matthews MW, Bernard S, Robertson L. An algorithm for detecting trophic status (chlorophyll-a), cyanobacterial-dominance, surface scums and floating vegetation in inland and coastal waters. Remote Sensing of Environment. 2012;**124**:637-652

[56] Lunetta RS, Schaeffer BA, Stumpf RP, Keith D, Jacobs SA, Murphy MS. Evaluation of cyanobacteria cell count detection derived from MERIS imagery across the eastern USA. Remote Sensing of Environment. 2015;**157**:24-34

[57] Nazeer M, Wong MS, Nichol JE. A new approach for the estimation of phytoplankton cell counts associated with algal blooms. Science of the Total Environment. 2017;**590**:125-138

[58] Myint S, Walker N. Quantification of surface suspended sediments along a river dominated coast with NOAA AVHRR and SeaWiFS measurements: Louisiana, USA. International Journal of Remote Sensing. 2002;**23**:3229-3249

[59] Wass P, Marks S, Finch J, Leeks GJL, Ingram J. Monitoring and preliminary interpretation of in-river turbidity and remote sensed imagery for suspended sediment transport studies in the Humber catchment. Science of the Total Environment. 1997;**194**:263-283

[60] V. e. NASA. Modis: Mouth of the Yangtze. 2018. Available from: https://visibleearth.nasa.gov/view.php?id=55219

[61] Copernicus E. Copernicus Open Access Hub. 2018. Available from: https://scihub.copernicus.eu/dhus/#/home

[62] Nechad B, Ruddick K, Park Y. Calibration and validation of a generic multisensor algorithm for mapping of total suspended matter in turbid waters. Remote Sensing of Environment. 2010;**114**:854-866

[63] Novo E, Hansom J, Curran P. The effect of viewing geometry and wavelength on the relationship between reflectance and suspended sediment concentration. International Journal of Remote Sensing. 1989;**10**:1357-1372

[64] Curran P, Hansom J, Plummer S, Pedley M. Multispectral remote sensing of nearshore suspended sediments: A pilot study. International Journal of Remote Sensing. 1987;**8**:103-112

[65] Feng L, Hu C, Chen X, Song Q. Influence of the Three Gorges Dam on total suspended matters in the Yangtze Estuary and its adjacent coastal waters: Observations from MODIS. Remote Sensing of Environment. 2014;**140**:779-788

[66] Doxaran D, Froidefond J-M, Lavender S, Castaing P. Spectral signature of highly turbid waters: Application with SPOT data to quantify suspended particulate matter concentrations. Remote Sensing of Environment. 2002;**81**:149-161

[67] Tian L, Wai OW, Chen X, Liu Y, Feng L, Li J, et al. Assessment of total suspended sediment distribution under varying tidal conditions in deep bay: Initial results from HJ-1A/1B satellite CCD images. Remote Sensing. 2014;**6**:9911-9929

[68] Nazeer M, Nichol JE. Combining landsat TM/ETM+ and HJ-1 A/B CCD sensors for monitoring coastal water quality in Hong Kong. IEEE Geoscience and Remote Sensing Letters. 2015;**12**:1898-1902

[69] Baban SM. Detecting water quality parameters in the Norfolk Broads, UK, using Landsat imagery. International Journal of Remote Sensing. 1993;**14**:1247-1267

[70] Khorram S, Cheshire H, Geraci AL, ROSA GL. Water quality mapping of Augusta Bay, Italy from Landsat-TM data. International Journal of Remote Sensing. 1991;**12**:803-808

[71] Gholizadeh M, Melesse A. Study on spatiotemporal variability of water quality parameters in Florida Bay using remote sensing. Journal of Remote Sensing and GIS. 2017;**6**:2

[72] Son S, Kim YH, Kwon J-I, Kim H-C, Park K-S. Characterization of spatial and temporal variation of suspended sediments in the Yellow and East China Seas using satellite ocean color data. GIScience & Remote Sensing. 2014;**51**:212-226

[73] Nechad B, Ruddick K, Neukermans G. Calibration and validation of a generic multisensor algorithm for mapping of turbidity in coastal waters. Remote Sensing of the Ocean, Sea Ice, and Large Water Regions. 2009;**2009**:74730H

[74] DiGiacomo PM, Washburn L, Holt B, Jones BH. Coastal pollution hazards in southern California observed by SAR imagery: Stormwater plumes, wastewater plumes, and natural hydrocarbon seeps. Marine Pollution Bulletin. 2004;**49**:1013-1024

[75] Holt B, Trinh R, Gierach MM. Stormwater runoff plumes in the Southern California Bight: A comparison study with SAR and MODIS imagery. Marine Pollution Bulletin. 2017;**118**:141-154

[76] Daling PS, Brandvik PJ, Mackay D, Johansen O. Characterization of crude oils for environmental purposes. Oil and Chemical Pollution. 1990;**7**: 199-224

[77] Shirvany R, Chabert M, Tourneret J-Y. Ship and oil-spill detection using the degree of polarization in linear and hybrid/compact dual-pol SAR. IEEE Journal of Selected Topics in Applied Earth Observations and Remote Sensing. 2012;**5**:885-892

[78] Korshenko A. Oil Spill Accident in the Kerch Strait in November 2007. Moscow, Russia: Nauka; 2011

[79] Carracedo P, Torres-López S, Barreiro M, Montero P, Balseiro C, Penabad E, et al. Improvement of pollutant drift forecast system applied to the Prestige oil spills in Galicia Coast (NW of Spain): Development of an operational system. Marine Pollution Bulletin. 2006;**53**:350-360

[80] Kim D-J, Moon WM, Kim Y-S. Application of TerraSAR-X data for emergent oil-spill monitoring. IEEE Transactions on Geoscience and Remote Sensing. 2010;**48**:852-863

[81] Rochman CM, Browne MA, Underwood A, Franeker JA, Thompson RC, Amaral-Zettler LA. The ecological impacts of marine debris: Unraveling the demonstrated evidence from what is perceived. Ecology. 2016;**97**:302-312

[82] Jambeck JR, Geyer R, Wilcox C, Siegler TR, Perryman M, Andrady A, et al. Plastic waste inputs from land into the ocean. Science. 2015;**347**:768-771

[83] Andrady AL, Neal MA. Applications and societal benefits of plastics. Philosophical Transactions of the Royal Society, B: Biological Sciences. 2009;**364**:1977-1984

[84] Andrady AL. Persistence of plastic litter in the oceans. In: Marine Anthropogenic Litter. Cham: Springer; 2015. pp. 57-72

[85] Eriksen M, Lebreton LC, Carson HS, Thiel M, Moore CJ, Borerro JC, et al. Plastic pollution in the world's oceans: More than 5 trillion plastic pieces weighing over 250,000 tons afloat at sea. PLoS One. 2014;**9**:e111913

[86] Maximenko N, Arvesen J, Asner G, Carlton J, Castrence M, Centurioni L, et al. Remote sensing of marine debris to study dynamics, balances and trends. In: Community White Paper Produced at the Workshop on Mission Concepts for Marine Debris Sensing. 2016

[87] Goddijn-Murphy L, Peters S, van Sebille E, James NA, Gibb S. Concept for a hyperspectral remote sensing algorithm for floating marine macro plastics. Marine Pollution Bulletin. 2018;**126**:255-262

[88] Serranti S, Palmieri R, Bonifazi G, Cózar A. Characterization of microplastic litter from oceans by an innovative approach based on hyperspectral imaging. Waste Management. 2018;**76**:117-125

[89] Moy K, Neilson B, Chung A, Meadows A, Castrence M, Ambagis S, et al. Mapping coastal marine debris using aerial imagery and spatial analysis. Marine Pollution Bulletin. 2018;**132**:52-59

[90] Martin C, Parkes S, Zhang Q, Zhang X, McCabe MF, Duarte CM. Use of unmanned aerial vehicles for efficient beach litter monitoring. Marine Pollution Bulletin. 2018;**131**:662-673

Chapter 3

The Hazards of Monitoring Ecosystem Ocean Health in the Gulf of Mexico: A Mexican Perspective

Luis A. Soto, Alejandro Estradas-Romero, Diana L. Salcedo, Alfonso V. Botello and Guadalupe Ponce-Vélez

Abstract

Ecological services provided by the Gulf of Mexico constitute vital assets for the socioeconomic development of the USA, Mexico, and Cuba. This ecosystem houses vast biodiversity and significant fossil fuel reserves. However, its ecological stability and resilience have been jeopardized by anthropogenic disturbances. Massive oil spills (Ixtoc-I, 1979; Deepwater Horizon, 2010) caused severe environmental injuries and unveiled the vulnerability of coastal and deep-sea habitats. Baseline and monitoring studies are actions implemented by the Gulf stakeholders to cope with such disturbances. The 3-year monitoring program implemented by Mexico in 2010 to assess the environmental damage caused by the Deepwater Horizon (DWH) event confirmed the void of knowledge on the complexity of physical and biological processes susceptible of being altered by oil spills. Between the pelagic and benthic compartments, the latter proved to be a better option in establishing the baseline concentration and trends of oil compounds. Surficial sediments exhibited an increasing concentration trend of PAH, AH, and trace metals throughout the 3-year monitoring. The macroinfauna and selected biomarkers experienced interannual variability attributed to critical hydrocarbon and trace metal thresholds. Sediment toxicity bioassays added support to the distribution and potential sources of oil contaminants dispersed from the northern gulf toward Mexican waters.

Keywords: Gulf of Mexico, oil spills, Deepwater Horizon, marine pollution, benthic ecology, macroinfauna

1. Introduction

Due to its geological origin, the Gulf of Mexico (GoM) represents an ideal semi-closed basin for the accumulation of fossil deposits of oil and gas [1]. This unique attribute has historically exposed the Gulf to natural seepage of oil and gas from the seabed. These natural emanations have been recorded in several sectors of the Gulf and represent a significant source of contamination [2, 3]. However, in recent times, the stability and resilience of this large marine ecosystem have been tested by severe anthropogenic disturbances. Massive spills of crude oil produced by the decontrol

IntechOpen

of the Ixtoc well-I in the Campeche Sound in 1979 and, most recently, in the well Macondo caused by the collapse of the Deepwater Horizon (DWH) oil platform off the coast of Louisiana in 2010 are examples that have caused severe damage to the environmental health of the Gulf [4, 5].

In April 20, 2010, a severe accident occurred at the oil platform DWH about 50 nautical miles southeast of the Mississippi River Delta, in the north of the GoM. This unfortunate event caused the loss of 11 lives and caused a spill of 4.9 million barrels of crude oil from the Macondo's well at 1650 m of depth. Several authors already considered this environmental catastrophe as the greatest disaster in the oil industry of the United States [6, 7].

The British firm British Petroleum (BP), responsible for the operation of the DWH platform, implemented a series of immediate emergency actions to mitigate somewhat the damage to the marine ecosystem, caused by the leakage of roughly 12,000–19,000 barrels of oil per day. Such activities involved the direct recovery of liquid hydrocarbons, the selective burning of oil slicks in surface waters, and the use of 1.85 million gallons of chemical dispersants (Corexit®), both on the surface and in the seabed [8, 9]. The hydrographic conditions prevailing in the GoM during the summer of 2010, combined with the onset of hurricane Alex in July, helped to contain the black tide of crude oil near the Mississippi Canyon. The initial oil slick trajectories were toward the northeastern sector of the Gulf. The satellite images obtained by the National Oceanic and Atmospheric Agency (NOAA), later supplemented with the ocean circulation models generated by the Consortium of Universities of the Northern Gulf, confirmed those trajectories, with ensuing filaments flowing toward the Texas coast. Similarly, through the use of remote sensors, it was possible to identify traces of crude oil trapped by the Eddy Franklin Gyre, a critical component of the loop current; some of these images also revealed small traces of crude oil in the waters of Mexico's exclusive economic zone (EEZ), reaching the north of the Yucatan Peninsula [10].

Without a doubt, the volume of crude oil spilled, and the quantity of chemical dispersant employed, constituted a severe alteration to the ecological balance and the environmental health of the GoM. The precise calculation of the volume of spilled oil, the trajectory of the oil stains in subsurface and surface waters, as well as the degradation rates of crude oil and its derivative compounds remain controversial topics among specialists. This problem is magnified by the chemical complexity of the crude oil. Fossil hydrocarbons include up to 17,000 organic compounds [11], each with its solubility, volatility, and density properties, as well as its different degrees of toxicity in marine biota and humans.

According to the American agencies, NOAA and the Environmental Protection Agency (EPA), a significant percentage of crude oil was recovered, and the rest was burned or lost by evaporation. However, there was overwhelming evidence of the severe damage caused to the coastal areas in the northeastern GoM. Marshlands, swamps, and coastal lagoons, which represent vital breeding grounds for wildlife fauna, were severely affected by the invading black tide. In the face of this dramatic environmental setting, Mexico was also forced to implement emergency measures focused on the early detection of crude oil slicks or tar balls within its vast exclusive economic zone (EEZ) in the GoM. Given the prevailing surface water circulation and the high connectivity among the different sectors of the Gulf, there was potentially a risk of oil pollutants from Macondo's entering Mexican waters. It was then necessary as a government to maintain a monitoring plan of the general oceanographic conditions in Mexican waters.

1.1 Research guidelines and observational strategy

Mexico, as a neighboring country of the USA, and Cuba, shares a vast ocean space in the GoM, bounded by 200 miles known as the EEZ. Under the international Treaty Law of the Sea, coastal countries are held accountable for the preservation and study of biotic and natural resources such as minerals, contained in both its waters as in the marine seabed. Based on this precept, and by the seriousness that represented the spill of fossil hydrocarbons introduced to the marine ecosystem of the GoM, it was imperative to implement a program of systematized oceanographic observations. Such program would contribute to build a dependable database of environmental parameters and thus carry out an assessment of environmental damage in the short term and midterm. Under the rules of international law [12], Mexico is obliged to have reliable information on the sources and the kinds of contaminants to assess the physical damage to coastal and ocean ecosystems in the GoM.

This chapter presents to the reader a synthesis of the most outstanding features of a 3-year research program of oceanographic observations (MARZEE) on the continental and upper slope off the coasts of the states of Tamaulipas and Veracruz during the period of 2010–2012. Considering the early dispersion forecasts of crude oil leaks originating from the north of the GoM, there was a high risk that the coast of the above states would be impacted by crude oil, preferably in the winter. To anticipate this potential anthropic disturbance, a monitoring program was implemented whose observational strategy included the sampling of 35 abiotic and biotic variables. Water, sediment, and biota from the continental shelf (50–183 m) and upper slope (200–>2000 m) were obtained in the summer of 2010 (M-I), and the winter of 2011 (M-II), and 2012 (M-III).

1.2 Focus of the research

- Water column structure (thermohaline profile, density, oxygen concentration, and fluorometry)
- Concentration of polycyclic aromatic hydrocarbons (PAHs) and aliphatic hydrocarbons (AHs) in the water column and surficial sediments
- Identification of potential pelagic and benthic bioindicator species
- Concentration of trace metals derived from oil compounds in surficial sediments
- Biomass, abundance, and density of phytoplankton, zooplankton, and benthos (macroinfauna)
- Stable carbon isotope values in and surficial sediments
- Sediment toxicity
- Concentration of trace metals derived from oil compounds in tissues of demersal fauna

List of authors and contributors.

Field	Principal researcher	Assistants
Hydrography	Jorge Zavala Hidalgo	A. Ruiz-Angulo, R. Romero-Centeno, O. Díaz-García, A. Contreras Ruiz-Esparza, M. Prospero-Díaz, E. Olvera Prado, N. Taylor-Espinoza
Aquatic biogeochemistry	Martin Merino Ibarra	V. Carnero-Bravo, S. Castillo, M. Pérez-Ramírez, M. Valdespino-Castillo
Marine pollution	Alfonso Vázquez Botello	G. Ponce-Vélez, C. García-Ruelas, S. Villanueva-Fragoso, F. Rivera-Ramírez, A. Montes-Nava, G. Díaz-González, M. Morales-Villafuerte
Phytoplankton	Sergio Licea Durán	R. Luna-Soria, P. Soto-Cadena, J. González-Fernández, M. Zamudio-Resendiz
Zooplankton	Laura Sanvicente Añorve	E. Lemus-Santana, G. Giles-Pérez, K. Arvizu-Coyotzi
Benthic ecology	Luis A. Soto	A. Estradas-Romero, D. Salcedo, K. Arvizu-Coyotzi, R. Aguilar-Escobar, M. Tapia-Domínguez, D. Chávez-Macedo, A. Flores-Celedón, R. David-Ávila, R. Galván-Bazán, C. Illescas-Monterroso, J. Ilhuicatzi-Torres
Ecotoxicology (bioassays)	Alma S. Sobrino Figueroa	
Ecotoxicology (trace metals)	Gabriel Núñez Noriega	

2. Materials and variables of database

As a result of three oceanographic campaigns MARZEE-I (summer, 2010), MARZEE-II (winter, 2011), and MARZEE-III (winter, 2012), aboard the R/V "Justo Sierra," a total of 93 oceanographic stations were sampled (35, 25, and 33, respectively) (**Figure 1**). At each station, water samples were collected at preselected depths (5, 10, 20, 30, 50, 75, 100, 150, and 200 m) in sites where high fluorescence (chlorophyll-*a*) was detected. A rosette device equipped with a conductivity, temperature, and depth sensors (CTD), 10 L Niskin bottles were deployed at each site. Zooplankton was sampled with a double bongo net, and benthos and sediments were obtained with either a Smith McIntyre grab or a Reineck box corer at depths ≤200 and >200, at the points of intersection of the isobaths of 50, 100, 200, 500, and 1500 (>2000 m only for MARZEE-III).

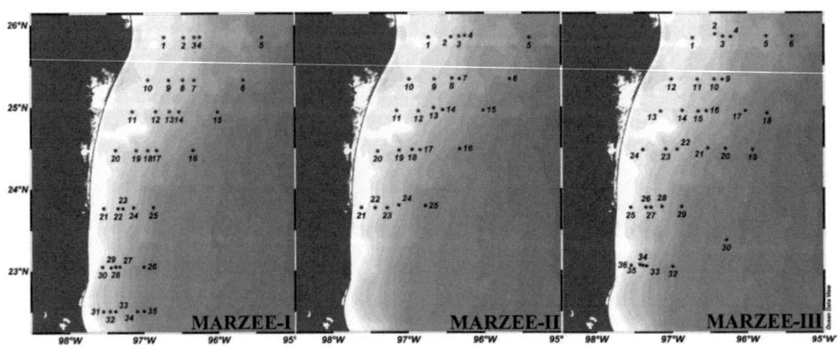

Figure 1.
Sampling sites location corresponding to the three oceanographic campaigns: MARZEE-I (M-I), MARZEE-II (M-II), and MARZEE-III (M-III).

We analyzed a total of 35 variables, distributed in the following manner:

- Water column: depth (m), salinity (UPS), temperature (°C), dissolved oxygen (mL/L), Brunt-Väisälä frequency (cycles/s), chlorophyll-*a* (μg/L), nitrates (μM), nitrites (μM), ammonium (μM), phosphates (μM), silicates (μM), PAH (μg/g), and AH (μg/g)

- Sediments: depth (m), sediment texture, organic carbon (%), inorganic carbon (%), terrestrial carbon (%), organic matter (%), $\delta^{13}C_{VDPB}$ (‰), PAH (μg/g), AH (μg/g), cobalt (μg/g), chromium (μg/g), nickel (μg/g), and vanadium (μg/g)

- Biota: phytoplankton (cells/L) and zooplankton (mL/100 m^3) abundance, zooplankton biomass (g/100 m^3), mortality (%), macroinfauna biomass (g C/cm^2), macroinfauna density (ind/10 cm^2), and induction factor SOS (SOSIF)

- Other: longitude (W) and latitude (N)

2.1 Data processing

To test whether there were significant interannual differences among the three sampling periods, and spatial differences in the study area, the environmental and biotic parameters were assessed using an analysis of variance based on permutations PERMANOVA. No transformation was required on environmental data, while biotic data were log transformed. A principal component analysis (PCA) was performed on environmental data. A nonmetric multidimensional scaling (nMDS) was conducted to analyze biotic data. A nonparametric BIO-ENV analysis [13] was also conducted to determine the relationship among environmental variables and biotic components. All these analyses were performed using PRIMER v6 & PERMANOVA *add on* statistical package [14, 15].

3. Study area

The GoM is one of the most diverse and productive world marine ecosystems. In this semi-closed basin, one can distinguish temperate, subtropical, and tropical habitats [16]. Its surface area is of approximately 1,768,000 km with a maximum depth of 4000 m in the central region [17, 18]. Mexico's EEZ in the Gulf has an extension of nearly 900,000 km^2, which represents 55% of its total surface area [19]. The area of study considered for this project is situated within Mexico's EEZ, on the northwestern corner of GoM. It spans from the northern end of the State of Tamaulipas, near the mouth of the Rio Bravo (approximately 26°N latitude), to the north of the State of Veracruz (22°N latitude) (**Figure 2**).

The study area has a surface area of 10,000 km^2, and its hydrographic conditions are highly influenced by the input of epicontinental, tropical, and subtropical marine waters [20]. The coastal zone of this part of the Gulf receives the runoff of several rivers (Bravo, Tuxpan, Pánuco, Indios Morales, Soto La Marina, and San Fernando or Carbonera). There are also two coastal lagoons systems: (a) the Laguna Madre, bounded on the north by the Rio Bravo's delta and on the south by the mouth of the Rio Soto la Marina and (b) the Laguna de Tamiahua, bounded on the north by the Pánuco River and on the south by the River Tuxpan [21, 22].

Its continental shelf lacks topographic irregularities. Its contour displays a gradual depth gradient ranging from 36 to 360 m. However, the floor of the

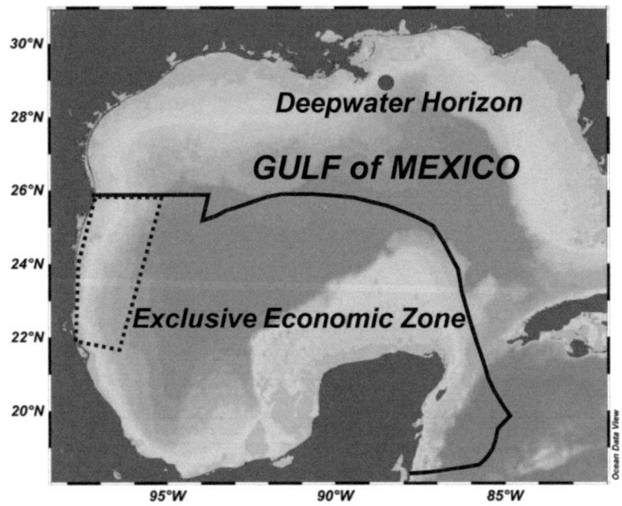

Figure 2.
Study area (dotted square). Exclusive economic zone (EEZ) (continuous line).

continental slope is rather abrupt, reaching depths between 540 and 1260 m. The continental shelf has a variable length—off the Rio Bravo reaches about 72–80 km, but toward the 23°N is close to 33–37 km—and further south just off Los Tuxtlas, Veracruz becomes narrower (between 6 and 16 km) [23].

The sea floor in the area of study is covered by muddy terrigenous sediments [24, 25] whose primary source is the sediment load discharged by the rivers mentioned above. The river runoff contributes to the formation of a strip of silty-sandy sediments running along the inner shelf. In the Tamaulipas coastal zone, sandy sediments prevail, while silts and clays are common far from the coast [17].

4. Hydrographic setting

The surface circulation of the GoM is dominated by the warm and saline waters that flow in through the Strait of Yucatan, forming the Loop Current (LC), and then exit at the Florida Strait [26]. In its passage through the Gulf Basin, anticyclonic gyres are formed from the LC, that later collide with the upper slope of the northwestern Gulf [27]. The speed of these vortexes (~6 km day^{-1}) and their residence time (~9–12 months) determine the distribution of physicochemical properties of the water masses, the circulation field, and the transport that controls the exchange of water masses between the continental shelf and the oceanic region [28, 29].

On the inner continental shelf on the west coast, in the province called "continental shelf and slope of the NW Gulf of Mexico" that goes from the south of Veracruz to the north of the Rio Bravo [30], the circulation is primarily toward the south from September to March and to the north from May to August. This circulation pattern produces temperature and salinity changes and coastal upwellings [31–33]. During the autumn and winter, cold fronts generated intense flows to the south that are alternated with periods of relative calm and flows to the north that coincide with high chlorophyll-*a* values at the surface. The summer-autumn conditions are less variable but are strongly affected by the passage of eddies and meteorological disturbances (tropical storm or hurricane); under these conditions, the lowest chlorophyll-*a* concentrations are recorded at the surface [26, 34]. Few are the studies on the dynamic conditions on the Tamaulipas coastline. The water masses on the platform are different from those on the slope or in the deep-sea. The circulation in the outer shelf and on the slope is often affected by the presence of cyclonic and anticyclonic eddies. When these are absent or weak, the circulation is toward the north. During the summer there is a semipermanent upwelling in the area, and during the winter there is advection of cold water and low salinity different to that of offshore waters [31, 32, 35]. On the slope, there is a strong influence of cyclonic and anticyclonic eddies generated in the east by the LC. These events do not have a seasonal periodicity or occur in the slope region. During the winter, the strong winds from the north (northerlies) maintain a homogenized water column, while in the summer the water column is stratified [26].

5. Oceanographic conditions

During M-I, toward the end of June, the studied area endured in its surface waters the effects of Hurricane Alex. The instability caused by this meteorological phenomenon produced strong turbulence in the water column along the coastal zone. Also, due to the unusual discharge from the Rio Bravo, salinity values were diluted in neritic waters, the concentrations of nutrients were high, and the zooplankton biomass exhibited a shift toward the north. There were abnormal values of oxygen and Chl-*a*. The interpretation of the hydrographic conditions and concentrations of nutrients indicated ascending conditions of the subsurface water in the northern sector and a sinking process of water in the southern sector. The upward motion of subsurface water took place mainly on the edge of the continental shelf, causing processes of fertilization in the euphotic zone.

During M-II, hydrographic conditions presented greater instability in the surface water, with a significant injection of water coming from the continental shelf of Louisiana to Texas. The structure of the water masses in the oceanic area was similar to the one described during M-I. The water column did not present a marked stratification in neritic waters, and the mixed layer was slightly deeper. On this occasion, no processes of upwelling of deep water were recognized nor intrusion of oceanic waters on the continental shelf.

The processes of convection of water masses that govern the concentrations of oxygen, Chl-*a*, and nutrients in the water column helped to maintain values of these variables within the normal ranges for neritic and oceanic waters of the GoM.

The concentrations of dissolved oxygen reported for the Gulf of Mexico vary from 2.4 to 5.4 mL/L [36, 37]. In this study, the oxygen remained relatively constant during the three oceanographic cruises, registering an average of 4.3 ± 0.8 mL/L. The highest values were recorded at the surface (<4.0 mL/L) and the lowest between 200 and 500 m (<3.0 mL/L). This layer corresponds to the Tropical Atlantic Central Water (TACW) located between 250 and 400 m. The minimum oxygen (2.6–2.9 mL/L) in M-II and M-III was recorded between 100 and 600 m depth.

In the case of the nutrients, the values showed a slight impoverishment (nitrates, 29.3–37.9 μM; silicate, 3.5–8.2 μM; phosphates, 1.9–3.4 μM). This fact emphasized the prevailing oligotrophic conditions (Chl-*a* < 0.25 ± 0.14 μg/L) reflected in the plankton components.

In M-III, the analysis of the density and the flotation frequency data, particularly at the isobaths of 500, 1500, and 2000 m, made it possible to distinguish the North Atlantic Subsurface Waters (NASW). Other identified water masses in the Gulf were as follows: North Atlantic Common Water (NACW), Tropical Atlantic Central Water (TACW), North Atlantic Intermediate Water (NAIW), and North Atlantic Deep Water (NADW) (**Figure 3**).

As indicated earlier, the region where intensive water mixing occurs is near the surface of coastal waters. The salinity and density values indicated the intrusion

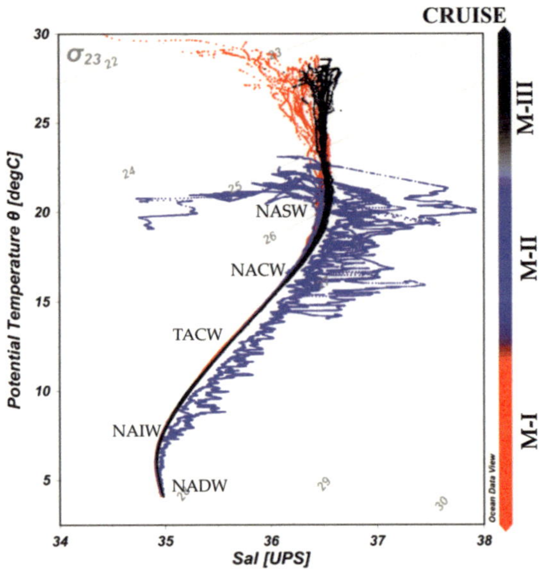

Figure 3.
T-S diagrams showing the profiles obtained during M-I (in red), M-II (in blue), and M-III (in black). North Atlantic subsurface waters (NASW), North Atlantic common water (NACW), tropical Atlantic central water (TACW); North Atlantic intermediate water (NAIW), and North Atlantic deep water (NADW).

of fresh water from the river discharge onto the Tamaulipas continental shelf. The degree of water mixing of the water column was estimated by calculating the Brunt-Väisälä frequency (N). The results of this procedure revealed a significant stratification in the three oceanographic campaigns (N > 0). The average value for each campaign was 6.6 cycles/s for M-I, 4.12 cycles/s for M-II, and 5.0 cycles/s for M-III. Furthermore, in M-I the highest N values corresponded to the depth of 30 m, which coincide with the thermocline's depth recorded between 10 and 35 m. In M-II, the highest N value was recorded at 75 m and in M-III at 50–70 m. In both instances, the thermocline depth was of 70–90 m and 50–70 m, respectively (**Figure 4**).

During M-II and M-III, the oxygen, nutrients (nitrates, phosphates, and silicates) and the Chl-*a* concentrations maintained values that fall within the known ranges reported for the GoM (0.05–2.5 μM PO_4, 0–35 μM NO_3; <0.29 ± 0.31 μg/L Chl-*a*) [36, 37].

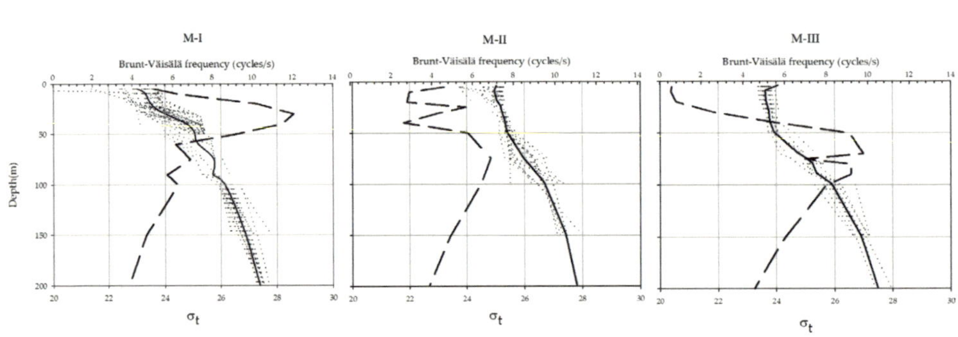

Figure 4.
Density and Brunt-Väisälä profiles for the three oceanographic cruises: M-I, M-II, and M-III. The dotted lines depict individual profiles, and the black line is the average. The long dash line is the Brunt-Väisälä profile associated with the density average.

6. Polycyclic aromatic hydrocarbons (PAHs)

The PAH included a range of 16 individual compounds that are commonly analyzed [38] in monitoring programs. The concentrations of total PAH in subsurface waters remained below the analytical detection limits (<0.003–0.03 μg/L) in the three surveys. Similarly, the ∑ individual PAH ranged between 0.1 and 0.02 μg/L. In M-I, benzo(a)anthracene, chrysene, benzo(b)fluoranthene, and indeno(1,2,3-cd)pyrene were recorded in 14.3% of the sites [39]. In M-II, only low concentrations of benzo(b)fluoranthene were recorded in seven sites which is equivalent to 28%. In W-III, the individual PAH identified were indeno(1,2,3-cd) pyrene, benzo(b)fluoranthene, benzo(a)pyrene, and pyrene. These compounds were detected in seven sites, representing 21.2%.

The concentrations of PAH contained in sediments fluctuated significantly among the three oceanographic campaigns. The values were similar to those previously recorded by [40] in the Tamaulipas continental shelf and by [41] in the Campeche and Tabasco continental shelves, regions widely exposed to intense oil activities. In M-I, the concentrations oscillated among 0.01 and 0.70 (0.29 ± 0.17 μg/g). These values decreased in 90% of the sampling stations in M-II, presumably as a consequence of the Hurricane Alex and associated rains. In this period, concentrations of 0.03–0.51 (0.16 ± 0.12 μg/g) were recorded. Throughout the following winter (M-III), PAH increased, presenting values of 0.05–1.54 (0.44 ± 0.03 μg/g). This increase indicated a recent deposit of hydrocarbons, considering that the sedimentation rate is very low in the deep GoM [42].

The total PAH recorded in sediments exhibited a heterogeneous spatial distribution, but high concentrations were frequently recorded in the northern transect of the study area. In M-I, the highest concentrations were observed at 100 and 500 m depth, in M-II at 500 and 1500 m depth and in M-III at >2000 m depth. The presence of high concentrations of PAH in deep sediments (>500 m) is not likely related to Rio Bravo runoff but rather to a far-field transport of hydrocarbons from other than local sources. The observed interannual heterogeneity (temporal/spatial) in the PAH concentrations in the NW Gulf can find an explanation in the geochemical processes acting upon different sources of hydrocarbon compounds. Indeed, one of such processes is the biodegradation of fossil fuels oil by oil-degrading bacteria [3, 43] that takes place when a massive oil spill occurs.

The distribution of individual PAH was heterogeneous throughout the study, but among the predominant were the chrysene in M-I (0.06 ± 0.02 μg/g), the fluorene in M-II (0.06 ± 0.01 μg/g), and the benzo(a)anthracene in M-III (0.08 ± 0.07 μg/g). These are low molecular weight compounds generated from the burning of fossil fuels and are abundant in crude oil, so they are indicators of a recent input of anthropogenic hydrocarbons [44]. The benzo(a)anthracene and chrysene have acute toxicity and represent an environmental risk.

The primary origin of PAH in the area of study was pyrolytic. In some stations, the combustion of fossil fuels is predominated, while in others, the combustion of organic carbon, plants, wood, and other vegetal compounds prevailed. A mixture of PAH of pyrolytic and petrogenic sources was restricted to few stations throughout the study. Among the petrogenic sources, crude oil was remarkable. This indicates an anthropogenic input to the area of study because of the fossil fuel burning.

Total PAH concentrations were below sedimentary quality criteria (LRE and MRE), whose exceeded limits indicate a potential adverse effect on benthic biota. However, among individual PAH, acenaphthylene, fluorene, benzo(a)anthracene, and dibenzo(a,h)anthracene exceeded the low-range effect (LRE) criterion in four stations in M-I. In M-III, acenaphthene, fluorene, dibenzo(a,h)anthracene, and anthracene exceeded the LRE in more than 23 stations. These results showed that the toxicity of the sediments caused by the presence of hydrocarbons increased throughout the study and hence the potential risk to the benthic fauna.

7. Trace metals

The concentrations of vanadium increased significantly over time in both the continental shelf and the continental slope (from 121.74 ± 14.44 μg/g in M-I to 144.86 ± 28.51 μg/g in M-III), showing a recent input. The values observed in this study in some deep regions of the northwestern Gulf may constitute evidence of the influence of the 1500 m depth plume oil derived from the accidental DWH oil spill. The concentrations of certain trace metals increase as the oil weathering increases [45]. Similarly, the concentration of nickel increased gradually and significantly over time (from 31 ± 4.87 μg/g in M-I to 42.16 ± 8.52 μg/g in M-III) exceeding the sediment quality criteria LRE in most of the stations throughout the study, particularly in sites deeper than 1500 m. The detection of values higher than the MRE during M-III indicated the potential damage to the benthic fauna. The concentration of this metal in deep sites may be linked to processes of sediment transport from the northern Gulf, which includes degraded petroleum products. The concentration of cobalt also increased slightly over time (from 12.51 ± 1.6 μg/g in M-I to 16.08 ± 2.61 μg/g in M-III). This trace element was mostly concentrated along the outer continental shelf and upper slope of the area of study, showing a similar pattern to that of vanadium and nickel. The chrome maintained dissimilar concentrations during the 3-year monitoring period of observation. The highest concentrations of Cr were detected in coastal areas exposed to the intensive river runoff from the Bravo River and Pánuco River.

8. Organic matter (OM) and stable carbon isotope ($\delta^{13}C$)

The inner continental shelf extended from Rio Soto La Marina River and the Laguna Madre represented vital deposition area of sedimentary organic matter of continental origin. The applied geochemical analysis revealed significant shifts in the concentrations of organic matter (OM) and organic carbon (OC) throughout the 3-year period of observation, showing a progressive increase over time. However, the values remained within the known ranges of concentrations previously recorded in the Gulf of Mexico. No significant changes were detected in the spatial pattern of distribution of organic inputs during the three periods of observation. However, the estimated OC percentages did show significant variability over time; such variability was more evident in deep sites (>1000 m), where presumably, there is a substantial accumulation of OM caused by processes of deposition or sediment transport on the continental slope.

The $\delta^{13}C$ values during the three campaigns fluctuated between −20.16 and −21.66‰ with an average of −21.02 ± 0.34‰. There seems to be and impoverished gradient of $\delta^{13}C$ values from the northwestern corner close to the coast, which gradually increased outward to the oceanic region, following a southeast pathway. The $\delta^{13}C$ results highlighted the predominance of autochthonous organic matter (marine) as the primary source of sedimentary OC over that of terrigenous origin, particularly at remote sites from the coast. However, as expected, near the coast, where there are important inputs of terrestrial organic matter derived primarily from C_4 plants, the OC isotopic signature is masked by the mixture with the autochthonous organic matter. In the present investigation, the isotopic fingerprint belonging to the Deepwater Horizon oil spill (−27.23 ± 0.03‰ for weathered petroleum and −27.34 ± 0.34‰ for crude oil) was not detected.

9. Ecotoxicology

9.1 Genotoxicity and bioassays

The short-term bioassays conducted to assess the toxicity of surface sediments, employing biomarkers, *Tetraselmis suecica* (microalgae), *Artemia franciscana* (crustacean), and *Brachionus plicatilis* (rotifers), revealed different responses (**Table 1**). The evaluation of toxicity using as indicator *T. suecica* showed a significant relationship with the presence of PAH, AH, and Fe. In contrast, *A. franciscana* showed distinct mortality pattern when exposed experimentally to the sediments from certain sites of the study area. For instance, significant toxicity was well-defined at sites on the continental shelf and slope just off the Rio Bravo, Soto La Marina, and Carrizales Rivers. The geochemical variables with the highest correlation with mortality of this species were the Zr and Rb during M-I, Ni in M-II, and PAH in M-III. The rotifer *B. plicatilis* served for the identification of areas of low and high toxicity through time. Not a defined pattern for both conditions was recognized, but there was clearly an increase in the temporary toxicity; during M-I, the high toxicity was reduced to two sites: one in front of Laguna Madre and another next to the Soto La Marina River. In the subsequent winter periods (2011–2012), the high toxicity initially corresponded to sites located between 500 and 1500 m of depth and then expanded onto the continental shelf. The mortality of this rotifer showed a significant correlation with the presence of Zr, Nb, Rb and SiO_2 in M-I, and PAH in both winter seasons.

Regarding the genotoxic effects of sediments analyzed, we were able to establish a significant correlation in the 3-year monitoring study, between the damage in the DNA structure and the concentration of PAH. Most of the stations with the highest levels of genotoxicity also presented the highest PAH concentrations. Statistically, we demonstrated an interannual decline of genotoxicity values. However, the percentage of sites containing sediments with substances fostering genotoxicity increased in M-III. The toxicity and genotoxicity are strongly linked to factors such as wastewater and industrial discharge and agrochemicals inputs in the coastal zone. However, in

Species	Toxicity level	M-I	M-II	M-III
Tetraselmis suecica	Low	17	20	6
	Medium	60	72	64
	High	23	8	30
Artemia franciscana	Low	6	92	6
	Medium	74	8	73
	High	20	0	21
Brachionus plicatilis	Low	29	24	6
	Medium	66	64	67
	High	5	12	27

Table 1.
Stations percentage by level of toxicity for each species, in the three MARZEE campaigns.

	Species	Vanadium		Nickel		Cadmium		Lead	
		Muscle	Liver	Muscle	Liver	Muscle	Liver	Muscle	Liver
M-I	*Cyclopsetta chittendeni*	0.19 ± 0.36	0.03 ± 0.028	2.24 ± 1.798	3.77	0.21 ± 0.484	1.44 ± 1.981	0.32 ± 0.363	1.24 ± 1.701
	Harengula pensacolae	0.13 ± 0.12		1.52 ± 0.019		0.02 ± 0.03		0.07 ± 0.034	
	Prionotus longispinosus	0.07 ± 0.02		4.40 ± 3.849		0.028 ± 0.033		0.1 ± 0.08	
	Pristipomoides aquilonaris	0.07 ± 0.007	1.72 ± 1.253	0.82 ± 0.319	4.34 ± 3.203	0.053 ± 0.053	1.54 ± 0.927	0.04 ± 0.02	0.37 ± 0.084
	Stenotomus caprinus	0.16 ± 0.25		3.7 ± 3.307		0.054 ± 0.049		0.09 ± 0.021	
	Synodus foetens	0.03 ± 0.01	0.03 ± 0.001	1.16 ± 0.566	3.65 ± 2.288	1.79 ± 3.052	8.73 ± 9.151	0.06 ± 0.03	0.21 ± 0.027
	Trachurus lathami	0.096		2.34		0.115		0.061	
	Centropristis philadelphica	0.01 ± 0.001	0.24 ± 0.145	0.46 ± 0.236	2.81 ± 1.344	0.01 ± 0.001	20.15 ± 15.798	0.03 ± 0.013	0.19 ± 0.042
M-II	*Cyclopsetta chittendeni*	0.03 ± 0.02		2.25 ± 1.463		0.01 ± 0.006		0.06 ± 0.05	
	Lutjanus synagris	0.01 ± 0.008	0.9 ± 0.026	1.37 ± 0.734		0.007 ± 0.003	0.19 ± 0.105	0.04 ± 0.009	1.38 ± 1.555
	Prionotus longispinosus	0.02 ± 0.005	0.52 ± 0.403	1.55 ± 1.17	16.69 ± 15.632	0.003 ± 0.004	26.68 ± 2.339	0.14 ± 0.079	1.14 ± 0.971
	Pristipomoides aquilonaris	0.02 ± 0.003	4.93 ± 3.51	0.46 ± 0.137	4.68 ± 2.968	0.006 ± 0.004	9.15 ± 7.653	0.02 ± 0.018	0.48 ± 0.379
	Synodus foetens	0.01 ± 0.009	0.15 ± 0.231	1.18 ± 0.904	0.38 ± 0.194	0.017 ± 0.006	3.02 ± 1.992	0.12 ± 0.11	0.96 ± 1.554
	Balistes capriscus	0.095	4.704	0.112	0.573		0.68	0.058	2.74
M-III	*Cyclopsetta chittendeni*	0.08 ± 0.04	0.80 ± 0.909	0.08 ± 0.035	0.67 ± 0.947	0.002 ± 0.002	0.082 ± 0.064	0.11 ± 0.067	1.33 ± 1.987
	Diplectrum bivittatum	0.10 ± 0.09	2.90 ± 4.472	0.14 ± 0.055	1.82 ± 1.928	0.004 ± 0.002	0.44 ± 1.019	0.15 ± 0.134	5.19 ± 9.038
	Lutjanus synagris	0.088	0.618	0.127	1.59		0.106	0.071	1.96
	Porichthys plectrodon	0.06 ± 0.05	0.89 ± 0.973	0.11 ± 0.094	0.30 ± 0.258	0.017 ± 0.014	0.65 ± 0.475	0.21 ± 0.186	0.48 ± 0.497
	Prionotus stearnsi	0.07 ± 0.008	4.39 ± 2.393	0.38 ± 0.438	1.09 ± 0.938	0.005	0.17 ± 0.095	0.11 ± 0.095	1.77 ± 0.722
	Prionotus spp	0.05 ± 0.01	0.31 ± 0.532	0.13 ± 0.025	0.18 ± 0.096	0.003	0.18 ± 0.341	0.09 ± 0.039	0.16 ± 0.085
	Raja texana	0.067	0.004	0.52	0.006		0.001	0.085	0.002
	Stenotomus caprinus	0.09 ± 0.07	0.25 ± 0.122	0.08 ± 0.02	0.51 ± 0.251		0.055 ± 0.026	0.023 ± 0.029	0.51 ± 0.447
	Synodus foetens	0.08 ± 0.06	3.06 ± 7.688	0.13 ± 0.106	1.48 ± 3.313	0.004 ± 0.003	0.30 ± 0.717	0.09 ± 0.072	1.66 ± 3.37

Table 2.
Average concentrations (μg/g) and standard deviation of the metals detected in muscle and liver tissues of fish in the three MARZEE campaigns.

deep zones (>500 m), the levels herein detected in both variables reflect the influence of different sources other than the regionals. Both the toxicity and genotoxicity of sediment can be attributed to the synergy between the PAH and other contaminants detected in sediments, including trace metals such as V, Ni, Cr, Co, Fe, and Al.

9.2 Trace metals in demersal fauna

The toxicity analysis of trace metals (vanadium, nickel, cadmium, and lead) in 250 tissues samples of demersal fauna (fish, crustaceans, and mollusks) showed that the vanadium was the metal less concentrated in the muscle tissue of fish. Concentrations of this metal showed variability over time, decreasing sequentially toward the M-III in muscle, and increasing in liver tissue, reflecting a null or low recent exposure to this metal. In contrasts, the nickel presented the highest concentration average values in liver tissue, in comparison with the muscle, throughout the three oceanographic campaigns. The high concentrations of Ni in some demersal fish may reflect its incorporation by benthic pray or sediment ingestion. The cadmium reached significant concentrations in the liver tissue of demersal fishes but lower concentrations in the muscle. Hence, according to the standard guidelines for human health, such concentrations did not pose any risk for direct consumption. The concentrations of lead recorded in muscle and liver tissues of fish did not exceed critical values of intake and therefore did not represent a risk for its consumption either (**Table 2**).

The vanadium appeared with a higher concentration in the muscle of macroinvertebrates (mollusks and crustaceans) (**Tables 3** and **4**). Nickel was also a persistent metal in macroinvertebrates, with highest concentrations at the end of the study (M-III). This metal showed different concentrations between crustaceans and mollusks, presumably due to their different capacities for bioaccumulation

	Species	Vanadium	Nickel	Cadmium	Lead
M-1	*Farfantepenaeus aztecus*	0.501 ± 0.528	0.095 ± 0.175	0.025 ± 0.071	0.026 ± 0.08
	Sicyonia sp.	1.013 ± 1.144	0.048 ± 0.059	0.001 ± 0.001	0.005 ± 0.004
	Squilla sp.	0.396 ± 0.233	0.064 ± 0.05	0.591 ± 0.393	0.006 ± 0.006
M-II	*Farfantepenaeus aztecus*	1.053 ± 0.747	0.006 ± 0.001		0.003 ± 0.001
	Squilla sp.	1.847 ± 0.164	0.022 ± 0.009	0.332 ± 0.206	0.002 ± 0.0008
M-III	*Farfantepenaeus aztecus*	0.19 ± 0.105	0.413 ± 0.371	0.043 ± 0.085	0.142 ± 0.116
	Rimapenaeus similis	0.199 ± 0.051	0.476 ± 0.101	0.014 ± 0.008	0.095 ± 0.069
	Sicyonia dorsalis	0.139	0.28		0.195
	Sicyonia typica	0.323 ± 0.238	1.828 ± 2.416	0.052 ± 0.070	0.623 ± 0.460
	Solenocera necopina	0.096 ± 0.057	0.170 ± 0.129	0.01	0.139 ± 0.142
	Solenocera vioscai	0.208	0.556	0.017	0.131
	Squilla chydaea	0.778 ± 0.405	1.402 ± 0.942	0.387 ± 0.069	0.863 ± 0.593

Table 3.
Average concentrations (μg/g) and standard deviation of the trace metals detected in crustaceans during the three MARZEE campaigns.

	Species	Vanadium	Nickel	Cadmium	Lead
M-I	*Loligo* sp.	0.149 ± 0.05	0.007 ± 0.005	0.039 ± 0.027	0.002 ± 0.001
	Amusium papyraceum	4.646	0.003	0.013	0.0001
	Mercenaria campechiensis	7.768	0.02	0.001	0.003
M-II	*Loligo* sp.	0.091 ± 0.084	0.013 ± 0.016	0.001 ± 0.001	0.0006 ± 0.0002
	Amusium papyraceum	7.251 ± 0.855	0.054 ± 0.0003	1.029 ± 0.019	0.004 ± 0.001
M-III	*Loligo* sp.	0.32 ± 0.271	0.408 ± 0.213	0.044 ± 0.023	0.384 ± 0.132

Table 4.
Average concentrations (μg/g) and standard deviation of the trace metals detected in mollusks during the three MARZEE campaigns.

and regulatory mechanisms of excretion. The cadmium represented the metal with lower concentrations in the tissue of macroinvertebrates. The recorded values of Cd did not exceed those established by the safety guidelines for human health. Only in the case of a crustacean predator (*Squilla* sp.), an average concentration of 0.592 ± 0.394 μg g^{-1} was registered during M-I, which exceeded the safety limits. In the case of lead, its concentration in the muscle of crustaceans and mollusks fit for human consumption remained below 1 μg/g, except for three species of crustaceans recorded in M-III. This value is considered as critical threshold for human health. In summary, the analysis conducted of metals in tissues of demersal fish, crustaceans, and mollusks did not indicate life-threatening concentrations for the individuals nor to the human health in most of the cases. However, one cannot overrule the possible existence of bioaccumulation and biomagnification phenomena that eventually might affect the demersal trophic web.

10. Plankton

10.1 Phytoplankton

According to the taxonomic composition and abundance of phytoplankton algae, it was found that the values obtained coincided with those previously reported for this region (**Table 5**) [46]. These results suggest oligotrophic conditions, as confirmed by the low nutrients (nitrates, 29.3–37.9 μM; silicate, 3.5–8.2 μM; phosphates, 1.9–3.4 μM) and chlorophyll-*a* concentrations (>0.25 ± 0.14 μg/L). The abundance of dinoflagellates and phytoflagellates, and the low diatom abundance in most of the analyzed samples, adds support to the oligotrophic condition of this region in the summer and winter seasons. The Chlorophyceae algae were responsible for the blooms recorded in coastal waters (652, 179 cells/L). No significant differences were found in abundance among the three campaigns.

10.2 Zooplankton

The zooplankton biomass values registered in the three oceanographic campaigns fluctuated between 1.20 and 19.38 g/100 m^3. These values were considered impoverished when compared to those registered in the SW Gulf, which exceed 40 and 100 g/100 m^3 [47]. In the two winter seasons (2011 and 2012), the

	M-I	M-II	M-III
Stations	35	25	33
Samples	282	188	230
Abundance (cell L^{-1})	35–19.6 × 10^4	12–7.75 × 10^4	66–66.5 × 10^4
Groups			
• Diatoms	145	104	89
• Dinoflagellates	175	62	58
• Silicoflagellates	16	3	2
• Cocolithophorids	8	1	0
• Cyanophytes	7	1	2
• Chlorophytes	2	1	1
• Ciliated	1	1	0
• Total species	354	173	152

Table 5.
Phytoplankton abundance by groups (cells/L) recorded in the water samples obtained during the three MARZEE campaigns.

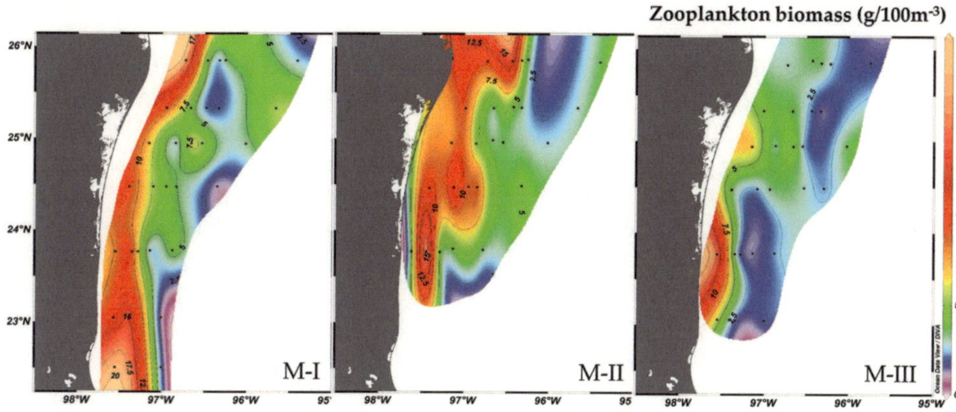

Figure 5.
Zooplankton biomass (g/100 m³) distribution for the three campaigns M-I, M-II, and M-III.

zooplankton revealed a significant decrease in biomass in both neritic and oceanic waters. In 2011 the biomass varied between 2.9 and 19 g/100 m^3, and in 2012 it reached 1.2–15.8 g/100 m^3. The neritic waters showed high variability in biomass (2–7 g/100 m^3), due to the influence of river discharges and the intrusion of ocean water near the coast. The zooplankton biomass was less than 8 g/100 m^3 in M-I and in M-II, while in M-III, it was less than 3 g/100 m^3 (**Figure 5**).

11. Infaunal benthic community

The taxonomic composition, density, and biomass of the infaunal benthic biota constituted a valuable analytical asset in the effort of identifying the magnitude of natural changes opposed to those potentially caused by anthropogenic disturbances.

In M-I, an impoverished infaunal benthic community was recorded, with only five taxa recorded and an average density value of 4.64 ± 7.03 individuals/10 cm^2. In M-II, the diversity of taxa continued being poor, recording seven taxa (**Table 6**), but

Taxa	M-I	M-II	M-III
Annelida	10.97	98.44	209.63
Arthropoda	9.20	17.70	48.15
Mollusca	40.01	4.24	16.99
Nematoda	101.27	60.19	155.80
Priapulida	0	0.70	4.72
Sipunculida	0	0.70	3.30
Nemertea	0	1.41	8.97
Kinorhyncha	1.06	0	3.77

Table 6.
Densities per taxon (ind/10 cm²) recorded in the three MARZEE campaigns.

the density values showed a small increase: 7.33 ± 8.48 individuals/10 cm^2. In M-III, the highest diversity and density values were recorded: eight taxa and 13.67 ± 22.71 individuals/10 cm^2, respectively [48]. The pattern of density in both seasons maintained the same negative exponential correlation with respect to the depth. Interestingly, there were significant density values at sites on the shelf rich in organic materials exported from the coastal zone; similar density values were also recorded in deeper sites in which presumably deposition and sediment transport occur. The nonmetric multidimensional scaling (nMDS) analysis applied to the estimated infaunal density in the three campaigns confirmed that M-I was different to the winter of 2011 and 2012; while the latter were similar to each other.

Significant temporal differences among the three campaigns were detected through the PERMANOVA analysis. Pairwise test indicated that such differences were interannual rather than seasonal. Spatially, only significant bathymetrical differences were detected; no latitudinal significant differences were noted. The pairwise test showed differences among the benthic infauna of the inner continental shelf (50 m) and deeper strata [48].

Based on the interpretation of abundance/biomass comparison curves (ABC) of the macroinfaunal community, it was possible to assess its interannual ecological equilibrium expressed as a stress factor. A clear trend of position of the curves since 2010–2012 revealed an interannual intensification of the stress degree. We inferred that the proliferation of nematodes in the latter season is symptomatic of such stress condition (**Figure 6**) [48].

In M-III, the infaunal community experienced a substantial change in its composition. The nematode worms reached a high dominance (44%). Even though no statistically significant latitudinal or bathymetric patterns of dispersion were distinguished, high density values were concentrated near the 50 m isobaths. The notorious abundance of the genus *Sabateria* in our samples deserves special attention. This genus represents an invaluable biomarker due to its tolerance to high concentrations of organic matter, degraded, heavy metals, and hydrocarbons [49]. *Sabateria* is known as an opportunistic nematode which, together with other infaunal dwellers like *Terschellingia*, *Paracomesoma*, and *Daptonema*, are normally found in highly contaminated sediments by organic matter characterized by a low redox potential [49, 50].

Metazoan organisms that make up the infaunal community are particularly sensitive to alterations in the geochemical properties of the sediments. A multivariate analysis BIO-ENV was performed to relate the set of environmental sedimentary variables to the macrofauna community structure. The correlation values obtained

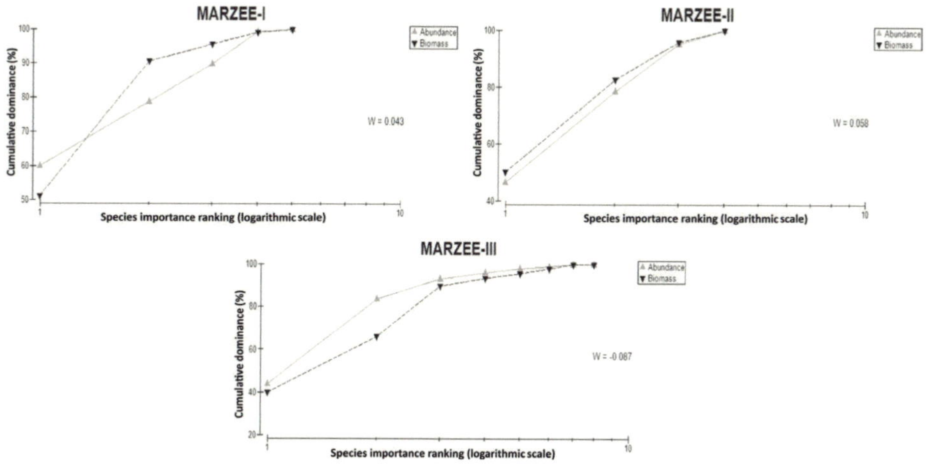

Figure 6.
Abundance/biomass comparison curves (ABC) for the three MARZEE campaigns.

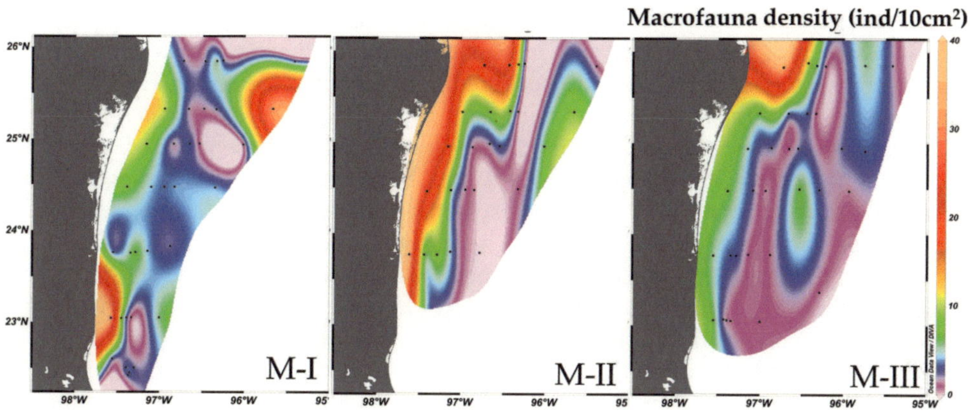

Figure 7.
Spatial distribution of macrofauna density values (individuals/10 cm²) in the three MARZEE campaigns.

from the BIO-ENV were rather low in the three campaigns (<0.4). However, it was possible to identify the geochemical variables that seem to govern the macroinfauna distribution in each season. In M-I the variables were percentage of sand, the concentration of Al and V, and the $\delta^{15}N$ values. In M-II, important variables were Al PAH, Ni, and AH. In M-III, the variables were Al, V, and AH. However, the correlation values were not significant. Nonetheless, in the summer season (2010), the influence of natural variables is more obvious than in the following periods (2011 and 2012), in which variables linked to crude oil become more relevant (**Figure 7**).

The changes observed in the community variables such as taxonomic composition and density of the macroinfaunal components were attributed to the gradual increase recorded in the study area of MO, HA, PAH, and metals such as Ni, V, and Co.

12. Final remarks

Many are the factors that determine the final destiny of complex oil molecules in the marine ecosystem. Coastal habitats (lagoons, coral; reefs, marsh; and lands, mangroves), open waters, and seabed are vulnerable to oil contaminants due to lasting effects of toxic compounds incorporated in the trophic web or deposited in shallow and deep sediments. Our research revolved around two major premises: (a) the existence of trans-boundary pollutants in the GoM and (b) the high connectivity of oceanographic processes within the GoM. For Mexico, these two concepts are essential in understanding the potential environmental consequences of a massive oil spill in its EEZ and shoreline. The above two conditions facilitate the active transport of contaminants across different sectors of the Gulf. There are no physical barriers or other sort of factors that impede the free passage from the US waters toward Mexico's EEZ and vice versa. Migratory species (trans-boundary species) and planktonic larvae take advantage of the ocean circulation to extend their fundamental niche (growth-reproduction-nutrition) within the Gulf, regardless of international legal boundaries.

During the 3-year monitoring program in Mexico's EEZ in the aftermath of a major oil spill in the northern Gulf of Mexico caused by the Deepwater Horizon event in April of 2010, our research efforts focused on the assessment of crude oil compounds in water, sediments, plankton, and benthos of the NW Gulf. The high-connectivity and the trans-boundary mechanisms that facilitate the dispersion of larvae and pollutants within the GoM were the essential premises in examining the far-field effects of the DWH oil spill in Mexican waters. Therefore, it was doubtful that the DWH harmful effects were confined to a restricted area near the Macondo's wellhead.

The information contained in this chapter is an excellent baseline environmental data from more than 35 hydrographic variables, biogeochemical and biological properties of the benthic and pelagic ecosystems of the continental shelf, and the upper slope of the NW Gulf of Mexico. The analyses and interpretation so far achieved in this first multidisciplinary effort do serve to recognize the significant alterations in the sedimentary quality standards and the risk of harmful effects on benthic organisms, attributable to anthropogenic factors.

The lack of knowledge on the long-term environmental damage compels us to implement innovative research approaches. If we consider the short-term monitoring operations, it is necessary to adjust the network of observation sites and optimize the number of variables focusing on the detection of toxic elements in water, sediment, and the trophic web. In a long-term scenario, it is essential to continue with multidisciplinary monitoring programs involving research vessels and stationary observational buoys.

Acknowledgements

The authors would like to express their gratitude to the members of the Benthic Ecology Laboratory for their assistance in the sample processing. A special word of appreciation to the crew of the R/V Justo Sierra for their invaluable support during the 3-year monitoring program: "Marco Ambiental de las Condiciones Oceanográficas en el Sector NW de la ZEE de México en el Golfo de México (MARZEE)." Funds for this project were provided by the Instituto Nacional de Ecología y Cambio Climático (INECC). A word of appreciation for the support provided by the Institute of Marine Sciences and Limnology (ICMyL, UNAM) for the development of this project.

Author details

Luis A. Soto[1*], Alejandro Estradas-Romero[2], Diana L. Salcedo[3], Alfonso V. Botello[4] and Guadalupe Ponce-Vélez[4]

1 Oceanic and Coastal Processes Academic Unit, Benthic Ecology Laboratory, Institute of Marine Sciences and Limnology, National Autonomous University of Mexico, Mexico City, Mexico

2 School of Sciences, National Autonomous University of Mexico, Mexico City, Mexico

3 Graduate Program in Marine Sciences and Limnology, Benthic Ecology Laboratory, Institute of Marine Sciences and Limnology, National Autonomous University of Mexico, Mexico City, Mexico

4 Oceanic and Coastal Processes Academic Unit, Marine Pollution Laboratory, Institute of Marine Sciences and Limnology, National Autonomous University of Mexico, Mexico City, Mexico

*Address all correspondence to: lasg@cmarl.unam.mx

IntechOpen

© 2018 The Author(s). Licensee IntechOpen. This chapter is distributed under the terms of the Creative Commons Attribution License (http://creativecommons.org/licenses/by/3.0), which permits unrestricted use, distribution, and reproduction in any medium, provided the original work is properly cited.

References

[1] Núñez J, Cruz-Roque D, Barrera-Nabor P. Estudio de las emanaciones naturales de hidrocarburos en la Sonda de Campeche y sus efectos en las plataformas marinas de PEMEX. In: Soto LA, González C, editors. PEMEX y la Salud Ambiental de la Sonda de Campeche. Battelle Memorial Institute, Instituto Mexicano del Petróleo, Universidad Autónoma Metropolitana. Mexico, DF: Universidad Nacional Autónoma de México; 2009. pp. 329-350. ISBN: 9789684890404 9684890400

[2] Wilson RD, Monogham PH, Osanik A, Price LC, Rogers MA. Natural marine oil seepage. Science. 1974;**184**:857-865. DOI: 10.1126/science.184.4139.857

[3] MacDonald IR, Garcia-Pineda O, Beet A, Daneshgar AS, Feng L, Graettinger G, et al. Natural and unnatural oil slicks in the Gulf of Mexico. Journal of Geophysical Research, Oceans. 2015;**120**:8364-8380. DOI: 10.1002/2015JC011062

[4] Sammarco PW, Kolian SR, Warby RAF, Bouldin JL, Subra WA, Porter SA. Distribution and concentrations of petroleum hydrocarbons associated with the BP/Deepwater horizon oil spill, Gulf of Mexico. Marine Pollution Bulletin. 2013;**73**(1):129-143. DOI: 10.1016/j.marpolbul.2013.05.029

[5] Soto LA, Botello AV, Licea-Durán S, Lizárraga-Partida ML, Yáñez-Arancibia A. The environmental legacy of the Ixtoc-I oil spill in Campeche sound, southwestern Gulf of Mexico. Frontiers in Marine Science. 2014;**1**:1-9. DOI: 10.3389/fmars.2014.00057

[6] Joye SB, MacDonald IR, Leifer I, Asper V. Magnitude and oxidation potential of hydrocarbon gases released from the BP oil well blowout. Nature Geoscience. 2011;**4**(3):160-164. DOI: 10.1038/NGEO1067

[7] McNutt MK, Camilli R, Crone TJ, Guthrie GD, Hsieh PA, Ryerson TB, et al. Review of flow rate estimates of the Deepwater horizon oil spill. Proceedings of the National Academy of Sciences. 2012;**109**(50):20260-20267. DOI: 10.1073/pnas.1112139108

[8] Lubchenco J, Mcnutt M, Lehr B, Sogge M, Miller M, Hammond S, et al. Deepwater Horizon Oil Budget: What Happened to the Oil? Vol. 1. Silver Spring, MD: Natl Ocean Atmos Adm; 2010. Available from: http://www.noaanews.noaa.gov/stories2010/PDFs/OilBudget [Accessed: June 16, 2018]

[9] National Science and Technology Council, Subcommittee on Ocean Science and Technology. Deep-Water Horizon Oil Spill Principal Investigator Workshop: Final Report. Washington, DC: National Science and Technology Council, Subcommittee on Ocean Science and Technology; 2012. 91 p. Available from: http://www.marine.usf.edu/conferences/fio/NSTC-SOST-PI-2011/documents/SOST_2011_DWH_Workshop_Final_Report [Accessed: June 10, 2018]

[10] Soto LA, Vázquez-Botello A. Legal issues and scientific constraints in the environmental assessment of the Deepwater horizon oil spill in Mexico exclusive economic zone (EEZ) in the Gulf of Mexico. International Journal of Geosciences. 2013;**4**(5):39-45. DOI: 10.4236/ijg.2013.45B007

[11] Bjorlykke K. Petroleum Geoscience: From Sedimentary Environments to Rock Physics. Berlin: Springer; 2010. 517 p. DOI: 10.1007/978-3-642-02332-3_1

[12] Edelman PS. The oil pollution act of 1990. Environmental Law Review. 1990;**8**(1):1. Available from: https://digitalcommons.pace.edu/pelr/vol8/iss1/1 [Accessed: June 15, 2018]

[13] Clarke KR, Ainsworth M. A method of linking multivariate community structure to environmental variables. Marine Ecology Progress Series. 1993;**92**(3):205-219. DOI: 10.3354/meps092205

[14] Clarke KR, Warwick RM. Change in Marine Communities: An Approach to Statistical Analysis and Interpretation. 2nd ed. Plymouth: PRIMER-E; 2001. p. 176. ISBN: 978-1855311404

[15] Anderson MJ, Gorley RN, Clarke K. PERMANOVA for PRIMER: Guide to Software and Statistical Methods. Plymouth, UK: PRIMER-E Ltd; 2008. 214 p

[16] Toledo A, Soto LA. El Gran Ecosistema del Golfo de México: Marco conceptual. In: Botello AV, Rendón von Osted JA, Benítez A, Gold-Bouchot G, editors. Golfo de México Contaminación e impacto ambiental: Diagnóstico y tendencias. Campeche, México: UAC, UNAM, ICMYL, CINVESTAV-Unidad Mérida; 2014. pp. 1-36. ISBN: 968-5722-37-4

[17] Pica Y, Ponce MG, Barrón ME. Golfo de México y Caribe Mexicano: Oceanografía geológica. In: De la Lanza E, editor. Oceanografía de Mares Mexicanos. AGT Editor. México, DF; 1991. pp. 3-30. ISSN: 0187-6376

[18] Rivera-Arriaga E, Borges-Souza G. El gran ecosistema marino del Golfo de México: Perspectivas para su manejo. Jaina Boletín Informativo. 2006;**16**(1):30-48. ISSN 01800-4700

[19] Vidal FV, Vidal VM, Rodríguez-Espinoza PF, Sambrano-Salgado L, Portilla-Casilla J, Rendon-Villalobos JR, et al. Gulf of Mexico circulation. Revista de la Sociedad Mexicana de Historia Natural. 1999;**49**:1-15. ISSN: 03707415

[20] Delgado-Blas VH. Distribución espacial y temporal de poliquetos (Polychaeta) bénticos de la plataforma continental de Tamaulipas, Golfo de Mexico. Revista de Biologia Tropical. 2001;**49**(1):141-147. ISSN 0034-7744

[21] Ayala-Castañares A, Cruz R, García-Cubas A, Segura LR. Síntesis de los conocimientos sobre la Geología Marina de la Laguna de Tamiahua, Veracruz, México. In: Ayala-Castañares A, Phleger FB, editors. Lagunas Costeras, Un Simposio Memorias del Simposio Internacional sobre Lagunas Costeras. México, DF: UNAM-UNESCO; 1969. pp. 39-48

[22] Contreras EF. Las Lagunas Costeras Mexicanas. México: Secretaria de Pesca; 1988. 253 p

[23] Silva GD, Campos J. Facies sedimentarias en la porción de plataforma continental entre Tampico, Tamaulipas y Veracruz. Secretaria de Marina. Dirección General de Oceanografía Naval. Geología Marina; 1986. pp. 32-62

[24] Bouma AH. Distribution of sediments and sedimentary structures in the Gulf of Mexico. In: Rezak R, Henry VJ, editors. Contribution on the Geological and Geophysical Oceanography of the Gulf of Mexico. Volume 3. Texas A & M University Oceanography Studies; 1972. pp. 35-65. ISBN: 0872013480

[25] Lecuanda CR, Ramos LF. Distribución de sedimentos en la parte sur del Golfo de México. Informe técnico 2. Laboratorio de Sedimentología. Instituto de Ciencias del Mar y Limnología, Universidad Nacional Autónoma de México; 1985. pp. 1-23

[26] Xue Z, He R, Fennel K, Cai WJ, Lohrenz S, Hopkinson C. Modeling ocean circulation and biogeochemical variability in the Gulf of Mexico. Biogeosciences. 2013;**10**(11):7219-7234. DOI: 10.5194/bg-10-7219-2013

[27] Vidal VM, Vidal FV. La importancia de los estudios regionales de circulación

oceánica en el Golfo de México. Revista de la Sociedad Mexicana de Historia Natural. 1997;**47**:191-200. ISSN: 03707415

[28] Vidal VMV, Vidal FV, Pérez-Molero JM. Collision of a loop current anticyclonic ring against the continental shelf slope of the western Gulf of Mexico. Journal of Geophysical Research, Oceans. 1992;**97**(C2): 2155-2172. DOI: 10.1029/91JC00486

[29] Vidal VMV, Vidal FV, Hernández AF, Meza E, Zambrano L. Winter water mass distributions in the western Gulf of Mexico affected by a colliding anticyclonic ring. Journal of Oceanography. 1994;**50**(5):559-588. DOI: 10.1007/BF02235424

[30] Antoine JW. Structure of the Gulf of Mexico. In: Rezak R, Henry VJ, editors. Contribution on the Geological and Geophysical Oceanography of the Gulf of Mexico, Volume 3. Houston, Texas: Texas A & M Univ., Oceanographic Studies. Gulf Publ. Company; 1972. pp. 1-34. ISBN: 0872013480

[31] Zavala-Hidalgo J. Seasonal circulation on the western shelf of the Gulf of Mexico using a high-resolution numerical model. Journal of Geophysical Research. 2003;**108**(C12):3389. DOI: 10.1029/2003JC001879

[32] Zavala-Hidalgo J, Gallegos-García A, Martínez-López B, Morey SL, O'Brien JJ. Seasonal upwelling on the Western and southern shelves of the Gulf of Mexico. Ocean Dynamics. 2006;**56**(3-4):333-338. DOI: 10.1007/s10236-006-0072-3

[33] Dubranna J, Pérez-Brunius P, López M, Candela J. Circulation over the continental shelf of the western and southwestern Gulf of Mexico. Journal of Geophysical Research, Oceans. 2011;**116**(8). DOI: 10.1029/2011JC007007

[34] Martínez-López B, Zavala-Hidalgo J. Seasonal and interannual variability of cross-shelf transports of chlorophyll in the Gulf of Mexico. Journal of Marine Systems. 2009;**77**(1-2):1-20. DOI: 10.1016/j.jmarsys.2008.10.002

[35] Fernández-Eguiarte A, Zavala HJ, Romero R. Circulación de invierno en la Plataforma de Tamaulipas y áreas adyacentes. IX Reunión Nacional SELPER-México. CD-ROM; 1998.

[36] Morrison JM, Merrel WJ. Property distributions and deep chemical measurements within the Western Gulf of Mexico. Journal of Geophysical Research. 1983;**88**(C4):2601-2608. DOI: 10.1029/JC088iC04p02601

[37] El-Sayed SZ, Sackett WM, Jeffrey LM, Fredericks AD, Saunders RP, Conger PS, et al. Chemistry, primary productivity and benthic algae in the Gulf of Mexico. In: Bushnell VC, editor. Serial Atlas of the Marine Environment. Folio 22. 1972. p. 29. ISSN: 0517-192X

[38] USEPA (United States Environmental Protection Agency). Ecological Toxicity Information [Internet]. 2009. Available from: www.epa.gov/region5/superfund/ecology/html/toxprofiles.htm [Accessed: June 01, 2018]

[39] Botello AV, Soto LA, Ponce-Vélez G, Susana Villanueva F. Baseline for PAHs and metals in NW Gulf of Mexico related to the Deepwater horizon oil spill. Estuarine, Coastal and Shelf Science. 2015;**156**(1):124-133. DOI: 10.1016/j.ecss.2014.11.010.

[40] Ponce-Vélez G, Botello AV, Díaz-González G. Organic and inorganic pollutants in marine sediments from northern and southern continental shelf of the Gulf of Mexico. International Journal of Environment and Pollution. 2006;**26**(1/2/3):295-311. DOI: 295-311. 10.1504/IJEP.2006.009113

[41] Ponce G, Botello AV. Niveles de hidrocarburos en el Golfo de México. In: Botello AV, Rendón-von Osten J, Gold-Bouchot G, Agraz-Hernández C, editors. Golfo de México, contaminación e impacto ambiental: Diagnóstico y Tendencias. Universidad Autónoma de Campeche. Universidad Nacional Autónoma de México. Campeche, México: Instituto Nacional de Ecología; 2005. pp. 269-298. ISBN 968-5722-37-4

[42] Wade TL, Soliman Y, Sweet ST, Wolff GA, Presley BJ. Trace elements and polycyclic aromatic hydrocarbons (PAHs) concentrations in deep Gulf of Mexico sediments. Deep-Sea Research Part II: Topical Studies in Oceanography. 2008;**55**(24-26):2585-2593. DOI: 10.1016/j.dsr2.2008.07.006

[43] Hazen TC, Dubinsky EA, DeSantis TZ, Andersen GL, Piceno YM, Singh N, et al. Deep-sea oil plume enriches indigenous. Science. 2010;**330**:204-208. DOI: 10.1126/science.1195979

[44] Botello AV, Calva BLG. Polycyclic aromatic hydrocarbons in sediments from Pueblo Viejo, Tamiahua, and Tampamachoco lagoons in the southern Gulf of Mexico. Bulletin of Environmental Contamination and Toxicology. 1998;**60**(1):96-103. ISSN: 0007-4861

[45] Liu Z, Liu J, Zhu Q, Wu W. The weathering of oil after the Deepwater horizon oil spill: Insights from the chemical composition of the oil from the sea surface, salt marshes and sediments. Environmental Research Letters. 2012;**7**(3):14. DOI: 10.1088/1748-9326/7/3/035302

[46] Zamudio Resendis ME. Hidrología y fitoplancton en una región costera del oeste del Golfo de México [thesis]. Universidad Nacional Autónoma de México; 1998

[47] Sanvicente-Añorve LE. Comunidades ictioplanctónicas en el suroeste del Golfo de México [thesis]. Universidad Nacional Autónoma de México, Instituto de Ciencias del Mar y Limnología; 1990

[48] Salcedo DL, Soto LA, Estradas-Romero A, Botello AV. Interannual variability of soft-bottom macrobenthic communities of the NW Gulf of Mexico in relationship to the Deepwater horizon oil spill. Marine Pollution Bulletin. 2017;**114**(2):987-994. DOI: 10.1016/j.marpolbul.2016.11.031

[49] Keller M. Structure des peuplements mèoibenthiques dans le secteur pollue par le rejet en mer de L'egout de Marseille. Annales de l'Institut Océanographique. 1986;**62**(1):13-36

[50] Moreno M, Vezzulli L, Marin V, Laconi P, Albertelli G, Fabiano M. The use of meiofauna diversity as an indicator of pollution in harbours. ICES Journal of Marine Science. 2008;**65**(8):1428-1435. DOI: 10.1093/icesjms/fsn116

Chapter 4

Sediment and Organisms as Marker for Metal Pollution

Ong Meng Chuan and Kamaruzzaman Yunus

Abstract

Pollution caused by metal elements has drawn increasing attention worldwide due to the increase of anthropogenic contaminants to the marine ecosystems. Pollution of the natural environment by metals is a serious problem because these elements are indestructible and most of them have toxic effects on living organisms, when they exceed a certain concentration. Sediments are widely used as geo-marker for monitoring and identifying the possible sources since sediment can act as sink for the pollutants. Most metals are bound in fine-grain fraction because of its high surface area-to-grain size ratio where they have a greater biological availability compared to those in larger fraction. Lying in the second trophic level in the aquatic ecosystem, shellfish species have long been known to accumulate both essential and non-essential metals. Many researchers have reported the potentiality of using mollusks, especially mussel and oyster species, as bioindicators or biomarkers for monitoring the metal contamination of the aquatic system.

Keywords: metal pollution, sediments, geo-marker, organism, bio-markers

1. Introduction

Recently, marine environment such as coastal and estuarine regions is contaminated by waste created by human activities containing elevated concentrations of nutrients, organic pollutants, trace metals, and radionuclide [1, 2]. Some of these chemicals are highly toxic and persistent, and these elements have a strong tendency to become concentrated in marine food webs once they enter this aquatic environment. The pollution of coastal zones near metropolitan areas, by these anthropogenic wastes, is due to the large coastal human population and the enormous amounts of sewage discharged into coastal waters [3–6]. The addition of waste products into rivers, estuaries, and wetland environment (**Figure 1**), especially those in industrial and population centers, has led to a significant increase in this pollutant level, especially metal contamination [7]. Accumulation of metals in surface sediments from industrial effluents and urban sewage discharged into the aquatic environment without proper treatment will easily be identified through metal spatial variations in sediments [8, 9].

Rivers can transport metals into the marine environment, and the amount of the chemical element input to the oceans depends on their levels in the river sediments, water, suspended particulate matter, and the exchange processes that occur in the estuaries [10]. With recent industrialization and human activities (**Figure 2**) that happen in the coastal region, these metals are continuing to be discharged to

Figure 1.
Wetland ecosystem in Malaysia. This ecosystem may be polluted by metal pollutants derived from human activities. Photo by Ong Meng Chuan.

Figure 2.
Example of human activities (fishery industry) in the Gulf of Morbihan, France. Photo by Ong Meng Chuan.

estuarine and coastal environment through rivers, runoff, and land-based point sources where the chemical elements are produced as a result of metal refinishing by-products.

Metal concentrations in harbor or estuarine sediments usually are high due to significant anthropogenic contaminant loading carried by the upstream of tributary rivers and settled down at this area [11, 12]. The sediments itself can serve as a metal pool that can release metals to the overlying water via natural or anthropogenic chemical and physical processes, causing potential adverse health effects to organisms that live at the ecosystems [13, 14]. Moreover, marine organisms can uptake these chemical elements, which in turn enhances the potential of some elements entering into the food chain. Therefore, metal contaminations are considered by scientists as an environmental problem today in both developing and developed countries throughout the world [15].

Metals accumulate in the sediments through complex physical and chemical adsorption mechanisms depending on the nature of the sediment matrix and the properties of the adsorbed compounds [16, 17]. Several processes had been identified for controlling the metal concentration in sediment, such as direct adsorption by small particle of clays, adsorption of hydrous ferric and manganic oxides which may also associate with clay fraction, adsorption of natural organic substances associated with inorganic particle, and precipitation as new solid phases [18, 19]. With this unique characteristic, sediments are usually used as geo-marker for monitoring and identifying the potential pollution sources in aquatic environment. These sediment analyses are an important tool for the determination of pollutants as they sink in the bottom through different chemical constituents and can reflect the pollutant proxy in the environment. In addition, the sediments act as a useful indicator of long- and medium-term metal flux in industrialized estuaries and rivers, and they help to improve management strategies as well as to assess the success of recent pollution controls [20].

More than 90% of the metal compound load in marine aquatic systems is bound to suspended particulate matter and sediments [21]. Therefore, sediments serve as a pool of metals that could be released to the overlying water from natural and anthropogenic processes such as bioturbation and dredging, resulting in potential adverse health effects toward surrounding organisms [22, 23]. Besides that, it is necessary to determine the metal contamination in estuarine ecosystem because this area is the most productive ecosystem which serves as feeding area, migration route, and nursery area of many juvenile and adult organisms from freshwater and marine water ecosystem. Due of these important to the ecosystem, effective remedial actions to minimize the pollution by metals need to be distinguished if pollution are expected occurs there [24].

2. Sediment as geo-marker for monitoring study

Marine sediments (**Figure 3**), including materials originating from the terrestrial inputs, as well as atmospheric deposition and autogenetic matter from the ocean itself, preserve a continuous record of regional and even global environmental changes, which can be employed in metal pollution evolution [25, 26]. Because of its unique characteristic, sediment always is considered as mirror of sedimentary environmental changes, which can reflect the biological, geodynamic, and geochemical processes of former conditions [27, 28]. On the other side, environmental changes are not only driven by natural forces but also by anthropogenic effects by human [29]. Some studies had concluded that the anthropogenic impacts on the environment have led to eutrophication process in coastal zone and offshore and the interaction of the natural force and human activities has exerted great effects on the whole environmental system[30].

Figure 3.
Sediment sample usually used by researchers as geo-marker for pollution study. Photo by Ong Meng Chuan.

Sediments can pick up metals due to several chemical process and normally will settle down in marine aquatic environment. Because of this characteristic, sediment can act as an appropriate indicator to monitor the metal pollution. In aquatic environment, these pollutants are originated from natural and anthropogenic sources in the same manner [31]; thus, scientists have difficulty to identify and classify the origin of these pollutants in the environment. Therefore, to overcome these obstacles, several scientists were using sediment fraction and characterized them into several sizes to normalize the metal concentration [31, 32]. The rationale applying this approach is normally metals are associated with fine-grain fraction because this fraction has larger surface area and higher cation exchange capacity that can enhance metal adsorption [33]. These fine sediments such as silt and clay with size less than 63 μm (**Figure 3**) are categorized as the most geochemically active fraction in the sediment. With this characteristic, this fraction is suitable to determine the potential pollution in the sediment (**Figure 4**).

Because of their large adsorption capabilities, fine-grain sediments represent a major repository for metals and a record of the temporal changes in contamination. Thus, they can be used for historical reconstruction. Although metals can occur naturally in marine environment due to their presence in local rocks, it is difficult to differentiate whether the source of the metals comes from anthropogenic or natural sources. Therefore, for better understanding about the metal behavior and distribution, it is important to distinguish between metals released from natural processes and those anthropogenic mainly introduced by human activities.

Marine sediments play a key role in the geochemical and biological processes of an estuarine ecosystem. In particular, these sediments act as sinks for toxic metals that enter the estuary. This sediment characteristic can regulate the concentration of these minerals and compounds in the water column [34]. Marine sediment also plays a very important role in the physicochemical and ecological dynamics of metals in marine aquatic ecosystems. The physicochemical nature of sediment-bound metals is important in the bioaccumulation of aquatic organisms such as fishes and shellfish.

Figure 4.
Fine-grain sediments have high surface area-to-grain size ratio which can accumulate more metals in the sediment. Photo by Ong Meng Chuan.

Figure 5.
Sediment core collected from mangrove ecosystem to study the metal proxy and sediment accumulation rate. Photo by Ong Meng Chuan.

Sediment quality has been recognized as an important and sensitive indicator or geo-marker of environmental pollution by various scientists [35, 36] since sediments can act as an important sink for various pollutants, such as metals that had been discharged into the environment [37, 38]. Besides acting as pollution indicator, sediments are also important in the remobilization process of contaminants in aquatic environment under favorable conditions through the interaction process between waste column and surface sediments. Due to this process, scientists had developed several comprehensive methods to identify and assess the sediment contamination mainly to protect the marine aquatic organisms [39].

Over the last few decades, the study of sediment cores has shown to be an excellent tool for establishing the effect of anthropogenic and natural processes on depositional environments. Meanwhile, sediment cores (**Figure 5**) can provide chronologies of contaminant concentrations and a record of the changes in concentration of chemical indicators in the environment. During the early 1960s, sediment profiles from depositional areas were used to trace human activity, witnessed by anthropogenic contamination like phosphorus [40], and later in the 1970s, it was possible to distinguish radioactive isotope inputs due to nuclear tests. Metal accumulation rates in sediment cores can reflect variations in metal inputs in a given system over long periods of time. Hence, the study of sediments core provides historical record of various influences on the aquatic system by indicating both natural background levels and the man-induced accumulation of metals over an extended period of time. In addition, the dating of sediment cores using radioactive traces like 210Pb [41] permitted the precise quantification of the history of the inputs in a system [42].

3. Assessment of metal pollution level

The absolute concentration of metals in marine sediments never indicates the degree of contamination coming from either natural or anthropogenic sources because of its grain-size distribution and mineralogy characteristic [43, 44]. Normalization of metal concentrations to grain sizes, specific surface area, and reactive surface phases such as Li and Al is a common technique to remove artifacts in the data due to differences in depositional environments [45–47]. This method allows researchers to compare the contamination level directly even if the samples were collected at different locations. The most common normalization technique used is enrichment factor (EF) where this technique uses common elements such as Al, Li, and Fe as normalizer and index of geoaccumulation (I_{geo}) or compares the normalized concentration to average crustal abundance data [47, 48].

In order to examine to sediment status, the determined element concentrations normally were compared to the published background concentrations. Literature data on average world shale or sediment cores or sediments from pristine such as undisturbed wetlands and non-industrialized regions were analyzed to establish the background values. However, to reduce the metal variability caused by the grain sizes and mineralogy of the sediments and to identify anomalous metal contribution, geochemical normalization has been used with various degrees of success by employing conservative elements [49, 50]. Researchers have proposed various elements as normalizer, and these elements have the potential for the environmental studies. Some of them are lithium, Li [51–53]; aluminum, Al [54, 55]; scandium, Sc [56]; cesium, Cs [57, 58]; cobalt, Co [59]; and thorium, Th [60, 61]. Among all proposed normalizers, conservative elements, Li and Al, have been widely applied in marine and coastal study [62–64].

The concentration of metals in marine sediments cannot indicate the degree of contamination coming from either natural or anthropogenic sources because of grain-size distribution and mineralogy [44, 65]. Normalization of metal concentrations to sediment size, specific surface area, and reactive surface phases such as Li and Al is a common technique to remove artifacts in the data due to differences in depositional environments [46, 66]. This allows for a direct comparison to be made between contaminant levels of samples taken from different locations.

Based on the researches by several geochemists [67, 68], if an EF value is between 0 and 1.5, it is suggested that the metals may be entirely from crustal materials or natural weathering processes. If an EF is greater than 1.5, it is suggested that a significant portion of metals has arisen from non-crustal sources or anthropogenic pollution [61, 69].

Another common approach to evaluate the metal pollution in sediments is the index of geoaccumulation (I_{geo}) introduced by Müller [70] in order to determine and define metal contamination in sediments by comparing current concentrations with the background levels. Similar to metal enrichment factor, I_{geo} can be used as a reference to estimate the extent of metal pollution in sediments. The I_{geo} value is calculated by using the following equation:

$$I_{geo} = \log_2 (C_n/1.5B_n) \quad (1)$$

where C_n is the measured concentration of the element (n) in the sediment and B_n is the geochemical background concentration of the element (n). Factor 1.5 is the correction of background matrix factor due to the lithogenic effects [70]. The upper continental crust values of the studied metals are the same as those used in the aforementioned enrichment factor calculation [71]. Müller [70] has distinguished seven classes of the I_{geo} from Class 0 to Class 6. The highest class (Class 6) reflects at least 100-fold environment above the background value.

Class	Value	Sediment quality
0	$I_{geo} \leq 0$	Practically uncontaminated
1	$0 < I_{geo} < 1$	Slightly contaminated
2	$1 < I_{geo} < 1$	Moderately contaminated

Class	Value	Sediment quality
3	$2 < I_{geo} < 1$	Moderately to heavily contaminated
4	$3 < I_{geo} < 1$	Heavily contaminated
5	$4 < I_{geo} < 1$	Heavily to extremely contaminated
6	$5 < I_{geo} < 1$	Extremely contaminated

Tomlinson et al. [72] elaborated that the application of pollution load index (PLI) provides a simple way in assessing marine and coastal sediment quality by metal pollution. This assessment is a quick tool in order to compare the pollution status of different places [73]. PLI represents the number of times by which the metal concentrations in the sediment exceed the background concentration and gives a summative indication of the overall level of metal toxicity in a particular sample or location [74, 75]. PLI can provide some understanding to the public of the surrounding area about the quality of a component of their environment and indicates the trend spatially and temporarily [76]. In addition, it also provides valuable information to the decision-makers toward a better management on the pollution level in the studied region.

PLI is obtained as contamination factor (CF). This CF is the quotient obtained by dividing the concentration of each metal with the background value of the metal. The PLI can be expressed from the following relation:

$$\mathrm{PLI} = (\mathrm{CF}_1 \times \mathrm{CF}_2 \times \mathrm{CF}_3 \times \mathrm{CF}_4 \times \mathrm{CF}_{n)})^{1/n} \quad (2)$$

where n is the number of metals studied and the CF is the contamination factor. The CF can be calculated from

$$\mathrm{CF} = (\text{Metal concentration in samples/Background metal concentration}) \quad (3)$$

The PLI value more than 1 can be categorized as polluted, whereas less than 1 indicates no pollution at the study area [77, 78].

4. Ecological risk assessment by sediment quality guidelines

Over the last two decades, a considerable amount of research effort has been put into investigating sediment toxic threshold levels [79, 80]. As a result there are now a number of international guidelines relating to toxic concentrations as determined by field and laboratory data. The work of Long et al. [79] on sediment quality guidelines (SQGs) provides a useful tool for screening sediment chemical data to identify pollutants of concern and priorities problem sites (x). In their study, the toxicity range of these chemical pollutants in the sediments was estimated from experimental studies in the laboratory, observation, and measurement of these parameters in the field. The finding of the work can estimate the level of two pollutants that have high chances to give impact in adverse biological effect of 10 and 50% of biota population.

Using this approach, scientists classified the toxicity of metals into effect range low (ERL) and effect range median (ERM) concentrations [79]. The concentration value between ERL and ERM represents the intermediate range in which this concentration can give an impact in 10–50% of the organism populations. ERL indicates the chemical pollutant can be considered to be of minimal or low concern, and the adverse effects toward organisms are infrequently observed (<10%

impact on organisms population) if the concentrations are below the ERL value. On the other hand, ERM indicate that if the concentration is above this level which the significant effect can be observed in 50% or more of the organism population considered to be toxic and of significant concern.

5. GIS application in environmental study

Nowadays, the rapid developments of computer technology and geographical information system (GIS) are receiving increasing interest in environmental geochemistry study [81]. This method is becoming popular nowadays in marine environmental pollution studies to graphically and digitally present the distribution of metals in marine environments by using GIS technique [82, 83]. The spatial interpolation methods of geometrical interpolation, trend surface analysis, and kriging method are commonly used [84]. This base chemometric approach was applied to investigate the spatial distribution patterns of metals in marine sediment and to identify spatial human impacts on global and local scales [85, 86].

GIS is a tool for decision-making, using information stored in a geographical form, in this case, in isopleth map form. Some researchers defined major requirements and function of GIS and mentioned spatial data handling tool for solving complex geographical problems [87, 88]. This GIS approach is increasingly used in environmental pollution studies because of its ability in spatial analysis and interpolation, and spatial interpolation utilizes measured points with known values to estimate an unknown value and to visualize the spatial patterns [89]. On the regional and national scales, the geochemical mapping of metals can be used as a tool for visualization which is enhanced by computer-aided modeling using GIS to make it easier to identify the possible locations of contaminated area. At present, joint using of GIS and chemometric approach mainly focuses on river estuary [90], soil [91], and nonpoint source identification [92].

6. Organisms as biomarker for monitoring study

Marine aquatic organisms can accumulate metals from various sources in their surrounding environment. The possible sources of these metals include sediments and soil erosion [93, 94], air depositions of dust and aerosol [93, 95], and discharges of wastewater [93, 94]. The accumulation of metals in marine aquatic organisms can pose a long-term burden on biogeochemical cycling in the ecosphere [96]. Once the metals enter the food chain, they may accumulate to dangerous levels and be harmful to human health.

Shellfish species which are laying at the second trophic level in the aquatic ecosystem have long been known to accumulate both essential and nonessential metals. Many researchers have reported the potentiality of using mollusks, especially mussel and oyster species, as bioindicators or biomarkers for monitoring the metal contamination of the aquatic system [97, 98]. Besides being a biomarker for marine pollution studies, these mollusk species have also been used in ecotoxicology and toxicity studies. Individual biomonitors respond differently to different sources of bioavailable chemical elements, for example, in the solution, in sediments, or in foods. In order to conclude a complete picture of total metal bioavailability in a marine habitat, it is necessary, therefore, to use a correct biomonitor that can reflect the metal bioavailability in all available potential sources [99]. Such comparative use of different biomonitors should allow identification of the particular source of the contaminant elements [100] (**Figure 6**).

Figure 6.
Shellfish (left, green mussel; right, oyster) are commonly used as biomonitor to study the pollution status.

Metal accumulation in marine aquatic organisms depended on several factors, including the environmental concentrations of metals in water and sediments; the species of organisms; and body size and age of the marine organisms. Different concentrations of metal can also be found in different organs (stomach, gill, muscle, tissue) in the same biological sample [101, 102]. However, scientists mainly focused on the general metal burden in shellfish species such as oyster and mussel and the potential major pathways for metal contaminant in the coastal environment.

7. Choice of biomonitors for environmental study

Aquatic organisms can transport pollutants and contaminants into, within, and out of the marine aquatic ecosystem. These organisms can ingest the pollutants via water and food and inhale them as they breathe and during feeding process [103]. When the pollutants enter the organism body, some contaminants can quickly pass through several organs; however, some may be absorbed and accumulated in organism tissues, particularly fatty tissues [104]. Certain contaminants such as mercury and PCBs are easily dissolve in organism fats and oils but do not dissolve in water. Due to the organism metabolism process, bioaccumulation process can be clearly seen in carnivorous animals in higher tropic of food chain, ranging from big organism such as fishes and to human [105].

The choice of a suitable biomonitor needs to consider the potential sources of metals to the organism. For example, sea grass not in contact with sediments, therefore, will take up metals from dissolved sources only [99]. Suspension feeders take up metals both directly from seawater and from the suspended particles collected during feeding. Thus, mussels, oysters, and barnacles are all candidates as suspension feeding biomonitors, and a careful choice will differentiate between suspended particles of different size ranges. As a generalization, sessile barnacles, but not stalked barnacles, have evolved micro-feeding, using the first thoracic legs to filter small suspended particles which would pass through the setae of the expanded cirral net formed by the more posterior thoracic legs [106].

Deposit feeding bivalves will reflect the bioavailability of metals in the surrounding water via respiratory currents but also metal bioavailability in newly deposited particles, for they suck up such particles via the inhalant siphon during feeding [107]. Some bivalves are protected by the shell from contact with the interstitial water of the sediment, a protection not offered, for example, to a sediment burrowing polychaete, the soft epidermis of which may be bathed directly by interstitial water with a redox potential possibly very different from that of the overlying water [108].

As concluded by monitoring scientists [109, 110], species to be chosen as biomonitors should fulfill several criteria such as:

i. Sedentary organism or those fixed in one spot

ii. Easy to identify the species

iii. Abundant

iv. Long-lived

v. Available for all the time

vi. Large enough to provide sufficient sample

vii. Resistant to handle the organisms' stress during test preparation

viii. Adapt to environmental variations in physicochemical parameters such as salinity and temperature

8. Organisms as laboratory testing organisms

Besides using the organisms as a biomarker for metal pollution studies in the field, mollusk species also have been used in ecotoxicology and toxicity studies in the laboratory. Several criteria had been set in order to choose suitable organisms as testing organisms. Despite that, in order to achieve the objectives, these testing organisms should fulfill several criteria as follows:

i. Organisms should be commercially important and sensitive to the environment.

ii. Organism must be easy to obtain and maintain in the laboratory.

iii. Biology, feeding behavior, and their characteristic of the organism must be known.

iv. Organism must be healthy and free from disease.

v. Organisms should be acclimatized for at least 2 weeks before use.

vi. Mortality of organism in control tank must be less than 10%. If more than 10% mortality, the testing should be repeated.

Acknowledgements

Authors wish to express their gratitude to the metallic element research group researcher that contributes to the chapter content. Also thanks to the School of Marine and Environmental Sciences for funding the group to run the project and Oceanography Laboratory, PPSMS, for providing the facilities during the laboratory analysis.

Conflict of interest

The authors certify that they have no conflict of interest during preparation of this chapter.

Notes/thanks/other declarations

Thanks to the School of Marine and Environmental Sciences, Universiti Malaysia Terengganu, that provided us the facilities to run our research project related to metal pollution.

Author details

Ong Meng Chuan[1*] and Kamaruzzaman Yunus[2]

1 School of Marine and Environmental Sciences, Universiti Malaysia Terengganu, Kuala Nerus, Terengganu, Malaysia

2 Kulliyyah of Science, International Islamic University Malaysia, Kuantan, Pahang, Malaysia

*Address all correspondence to: ong@umt.edu.my

IntechOpen

© 2019 The Author(s). Licensee IntechOpen. This chapter is distributed under the terms of the Creative Commons Attribution License (http://creativecommons.org/licenses/by/3.0), which permits unrestricted use, distribution, and reproduction in any medium, provided the original work is properly cited.

References

[1] Clark MW, Davies-McConchie F, McConchie D, Birch GF. Selective chemical extraction and grain size normalisation for environmental assessment of anoxic sediments: Validation of an integrated procedure. Science of the Total Environment. 2000;**258**(3):149-170

[2] Kennish MJ. Environmental threats and environmental future of estuaries. Environmental Conservation. 2000;**29**(1):78-107

[3] Bothner MH, Casso MA, Rendigs RR, Lamothe PJ. The effect of the new Massachustts Bay sewage outfall on the concentrations of metals and bacterial spores in nearby bottom and suspended sediments. Marine Pollution Bulletin. 2002;**44**(10):1063-1070

[4] Matthai C, Birth G, Bickford G. Anthropogenic trace metals in sediment and settling particulate matter on a high-energy continental shelf (Sydney, Australia). Marine Environmental Research. 2002;**54**(2):99-127

[5] Sadiq M. Metal contamination in sediments from a desalination plant effluent outfall area. The Science of the Total Environment. 2002;**287**(1-2):37-44

[6] Bay SM, Zeng EY, Lorenson TD, Tran K, Alexander C. Temporal and spatial distributions of contaminants in sediments of Santa Monica Bay, California. Marine Environmental Research. 2003;**56**(1-2):255-276

[7] Jayaprakash M, Srinivasalu S, Jonathan MP, Ram-Mohan V. A baseline study of physico-chemical parameters and trace metals in waters of Ennore Creek, Chennai, India. Marine Pollution Bulletin. 2005;**50**(5):583-589

[8] Baptista Neto JA, Smith BJ, McAllister JJ. Heavy metal concentrations in surface sediments in a nearshore environment, Jurujuba Sound, Southeast Brazil. Environmental Pollution. 2000;**109**(1):1-9

[9] Dauvalter V, Rognerud S. Heavy metal pollution in sediments of the Pasvik River drainage. Chemosphere. 2001;**42**(1):9-18

[10] Chester R, Kudoja WM, Thomas A, Towner J. Pollution reconnaissance in stream sediments using non-residual trace metals. Environmental Pollution. 1985;**10**(3):213-238

[11] Paetzel M, Nes G, Leifsen LO, Schrader H. Sediment pollution in the Vagen, Bergen harbour, Norway. Environmental Geology. 2003;**43**(4):476-483

[12] Muniz P, Danula E, Yannicelli B, Garcia-Alonso J, Medina G, Bicego MC. Assessment of contamination by heavy metals and petroleum hydrocarbons in sediments of Montevideo Harbour (Uruguay). Environment International. 2004;**29**(8):1019-1028

[13] Fatoki OS, Mathabatha S. An assessment of heavy metal pollution in the East London and Port Elizabeth Harbours. Water SA. 2001;**27**(2):233-240

[14] McCready S, Birch GF, Long ER. Metallic and organic contaminants in sediments of Sydney Harbour, Australia and Vicinity—A chemical dataset for evaluating sediment quality guidelines. Environment International. 2006;**32**(4):455-465

[15] Zhang LP, Ye X, Feng H, Jing YH, Ouyang T, Yu XT, et al. Heavy metal contamination in western Xiamen Bay sediments and its vicinity, China. Marine Pollution Bulletin. 2007;**54**(7):974-982

[16] Ankley GT, Lodge K, Call DJ, Balcer MD, Brooke LT, Cook PM, et al. Heavy metal concentration in surface sediments in a nearshore environment, Jurujuba Sound, Southeast Brazil. Environment Pollution. 1992;**109**(1):1-9

[17] Leivuori M. Heavy metal contamination in surface sediments in the Gulf of Finland and comparison with the Gulf of Bothnia. Chemosphere. 1998;**36**(1):43-59

[18] Gibbs RJ. Water chemistry of the Amazon River. Geochimica et Cosmochimica Acta. 1973;**36**(9):1006-1066

[19] Wen X, Allen HE. Mobilization of heavy metals from Le An River sediments. The Science of the Total Environment. 1999;**227**(2-3):101-108

[20] Ravichandran M, Baskaran M, Santschi PH, Bianchi TS. History of trace–metal pollution in Sabine-Neches Estuary, Beaumount, Texas. Environmental Science and Technology. 1995;**29**(6):1495-1503

[21] Calmano W, Hong J, Forstner U. Binding and mobilization of heavy metals in contaminated sediments affected by pH and redox potential. Water Science and Technology. 1993;**28**(8-9):223-235

[22] Daskalakis KD, O'connor TP. Distribution of chemical concentrations in US coastal and estuarine sediment. Marine Environmental Research. 1995;**40**(4):381-398

[23] Argese E, Bettiol C. Heavy metal partitioning in sediments from the lagoon of Venice (Italy). Toxicological and Environmental Chemistry. 2001;**79**(3-4):157-170

[24] Chapman PM, Wang F. Assessing sediment contamination in estuaries. Environmental Toxicology and Chemistry. 2001;**20**(1):3-22

[25] Wan GJ, Bai ZG, Qing H, Mather JD, Huang RG, Wang HR, et al. Geochemical records in recent sediments of Lake Erhai: Implications for environmental changes in low latitude–high altitude lake in Southwest China. Journal of Asian Earth Science. 2001;**21**(5):489-502

[26] Song JM. Biogeochemistry of China marginal seas. Shandong Press of Science and Technology. 2004:1-591

[27] Casado-Martinez MC, Buceta JL, Belzunce MJ, DelValls A. Using sediment quality guidelines for dredged material management in commercial ports from Spain. Environment International. 2006;**32**(3):388-396

[28] Dai JC, Song JM, Li XG, Yuan HM, Li N, Zheng GX. Environmental changes reflected by sedimentary geochemistry in recent hundred years of Jiaozhou Bay, North China. Environmental Pollution. 2007;**145**(3):656-667

[29] Kalis AJ, Merkt J, Wunderlich J. Environmental changes during the Holocene climatic optimum in Central Europe — Human impact and natural causes. Quaternary Science Reviews. 2003;**22**(1):33-79

[30] Hamed MA, Emara AM. Marine molluscs as biomonitors for heavy metal levels in the Gulf of Suez, Red Sea. Journal of Marine System. 2006;**60**(3-4):220-234

[31] Horowitz AJ, Rinella FA, Lamothe P, Miller TL, Edwars TK, Roche RL, et al. Variation in suspended sediment and associated trace element concentrations in selected riverine cross sections. Environmental Science and Technology. 1990;**24**(9):1313-1320

[32] Szefer P, Kusak A, Jankowska H, Wolowicz M, Ali AA. Distribution of

selected metals in sediment cores of Puck Bay, Baltic Sea. Marine Pollution Bulletin. 1995;**30**(9):615-618

[33] Horowitz AJ, Elrick K. The relation of stream sediments surface area, grain size and composition to trace element chemistry. Applied Geochemistry. 1987;**2**(4):437-451

[34] de Groot AJ, Salomons W, Allersma E. Processes affecting heavy metals in estuarine sediments. In: Burton JD, Liss PS, editors. Estuarine Chemistry. London: Academic Press; 1976. pp. 131-157

[35] Pekey H, Karakas D, Bakacoglu M. Source apportionment of trace metals in surface waters of a polluted stream using multivariate statistical analyses. Marine Pollution Bulletin. 2004;**49**(9-10):809-818

[36] Ong MC, Fok FM, Sultan K, Joseph B. Distribution of heavy metals and rare earth elements in the surface sediments of Penang River Estuary, Malaysia. Open Journal of Marine Science. 2016;**6**(1):79-92

[37] Tam NFY, Wong WS. Spatial variation of heavy metals in surface sediments of Hong Kong mangrove swamps. Environmental Pollution. 2000;**110**(2):195-205

[38] Bettinetti R, Giarei C, Provini A. A chemical analysis and sediment toxicity bioassays to assess the contamination of the River Lambro (Northern Italy). Archieves of Environmental Contamination and Toxicology. 2003;**45**(1):72-80

[39] Chapman PM. The sediment quality triad: Then, now and tomorrow. International Journal of Environment and Pollution. 2000;**13**(1-6):351-360

[40] Livingstone DA, Boykin JC. Vertical distribution of phosphorus in Linsley pond mud. Limnology Oceanography. 1962;7:57-63

[41] Robbins JA, Edgington DN. Determination of recent sedimentation rates in Lake Michigan using Pb–210 and Cs–137. Geochimica et Cosmochimica Acta. 1975;**39**(3):285-304

[42] Abrao JJ, Marques A, Bernat M, Wasserman JC, Lacerda LD. Metal Concentration in 210Pb Dated Sediment Profiles from a Sub-Tropical Brazilian Lagoon. Cartagena: International Symposium on Environmental Geochemistry in Tropical Countries; 1996. pp. 1-5

[43] Rubio B, Nombela MA, Vilas F. Geochemistry of major and trace elements in Ssediments of the Ria de Vigo (NW Spain): An assessment of metal pollution. Marine Pollution Bulletin. 2000;**40**(11):968-980

[44] Liu W, Li X, Shen Z, Wang D, Wai O, Li Y. Multivariate statistical study of heavy metal enrichment in sediments of the Pearl River Estuary. Environmental Pollution. 2003;**121**(3):377-388

[45] El Nemr A. Assessment of heavy metal pollution in surface muddy sediments of lake Burullus, southeastern Mediterranean, Egypt. Egyptian Journal of Aquatic Biology and Fisheries. 2003;**7**(4):67-90

[46] Santos IR, Silva-Filho EV, Schaefer CE, Albuquerque-Filho MR, Campos LS. Heavy metals contamination in coastal sediments and soils near the Brazilian Antarctic Station, King George Island. Marine Pollution Bulletin. 2005;**50**(2):185-194

[47] van der Weijden CH. Pitfalls of normalization of marine geochemical data using a common divisor. Marine Geology. 2002;**184**(3-4):167-187

[48] Cobelo-García A, Prego R. Heavy metal sedimentary record in a Galician Ria (NW Spain): Background values and recent contamination. Marine Pollution Bulletin. 2003;**46**(10):1253-1262

[49] Emmerson RHC, O'Reilly-Wiese SB, Macleod CL, Lester JN. A multivariate assessment of metal distribution in intertidal sediments of the Blackwater Estuary, UK. Marine Pollution Bulletin. 1997;**34**(11):484-491

[50] Lee CL, Fang MD, Hsieh MT. Characterization and distribution of metals in surficial sediments in Southwestern Taiwan. Marine Pollution Bulletin. 1998;**36**(6):464-471

[51] Loring DH. Lithium—A new approach for the granulometric normalization of trace metal data. Marine Chemistry. 1990;**29**:155-168

[52] Aloupi M, Angelidis MO. Normalization to lithium for the assessment of metal contamination in coastal sediment cores from the Aegean Sea, Greece. Marine Environment Research. 2001;**52**(1):1-12

[53] Soto-Jiménez MF, Paez-Osuna F. Distribution and normalization of heavy metal concentration in mangrove and lagoon sediments from Mazaltan (Gulf of California) Estuarine. Estuarine Coastal and Shelf Science. 2001;**53**(3):259-274

[54] Tuncel SG, Tugrul S, Topal T. A case study on trace metals in surface sediments and dissolved inorganic nutrients in surface water of Ölüdeniz Lagoon–Mediterranean, Turkey. Water Research. 2007;**41**(2):365-372

[55] Tessier E, Garnier C, Mullot JU, Lenoble V, Arnaud M, Raynaud M, et al. Study of the spatial and historical distribution of sediment inorganic contamination in the Toulon bay (France). Marine Pollution Bulletin. 2011;**62**(10):2075-2086

[56] Grousset FE, Quetel CR, Thomas B, Donard OFX, Lambert CE, Quillard F, et al. Anthropogenic vs lithogenic origins of trace element (As, Cd, Pb, Rb, Sb, Sc, Sn, Zn) in water column particles: Northwestern Mediterranean Sea. Marine Chemistry. 1995;**48**(3-4):291-310

[57] Ackerman F. A procedure for correcting grain size effect in heavy metal analysis of estuarine and coastal sediments. Environment Technology Letters. 1980;**1**(11):518-527

[58] Roussiez V, Ludwig W, Probst JL, Monaco A. Background levels of heavy metals in surficial sediments of the Gulf of Lions (NW Mediterranean): An approach based on 133Cs normalization and lead isotope measurements. Environmental Pollution. 2005;**138**(1):167-177

[59] Matthai C, Birch G. Detection of anthropogenic Cu, Pb and Zn in continental shelf sediments off Sydney, Australia—A new approach using normalization with cobalt. Marine Pollution Bulletin. 2001;**42**(11):1055-1063

[60] Larrose A, Coynel A, Schäfer J, Blanc G, Massé L, Maneux E. Assessing the current state of the Gironde estuary by mapping priority contaminant distribution and risk potential in surface sediment. Applied Geochemistry. 2010;**25**(12):1912-1923

[61] Strady E, Kervella S, Blanc G, Robert S, Stanisière JY, Coynel A, et al. Spatial and temporal variations in trace metal concentrations in surface sediments of the Marenne Oléron bay. Relation to hydrodynamic forcing. Continental Shelf Research. 2011;**31**(9):997-1007

[62] Din ZB. Use of aluminium to normalize heavy metal data from estuarine and coastal sediments of straits of Melaka. Marine Pollution Bulletin. 1992;**24**(10):484-491

[63] Tam NFY, Yao MWY. Normalization and heavy metal contamination in mangrove sediments. The Science of the Total Environment. 1998;**216**(1-2):33-39

[64] Schiff KC, Weisberg SB. Iron as a reference element for determining trace metal enrichment in Southern California coast shelf sediments. Marine Environmental Research. 1999;**48**(2):161-176

[65] Rubio B, Nombela MA, Vilas F. Geochemistry of major and trace elements in sediments of the Ria de Vigo (NW Spain): An assessment of metal pollution. Marine Pollution Bulletin. 2000;**40**(11):968-980

[66] Cobelo-García A, Prego R. Influence of point sources on trace metal contamination and distribution in a semi-enclosed industrial embayment: The Ferrol Ria (NW Spain). Estuarine, Coastal and Shelf Science. 2004;**60**(4):695-703

[67] Jiang FQ, Li AC. Geochemical characteristics and their implications to provenance and environment of surface sediments from the South Okinawa Trough. Acta Sedimentologic Sinica. 2002;**20**(4):680-686

[68] Zhang J, Liu CL. Riverine composition and estuarine geochemistry of particulate metals in China weathering features, anthropogenic impact and chemical fluxes. Estuarine, Coastal and Shelf Science. 2002;**54**(6):1051-1070

[69] Feng X, Li G, Qiu G. A preliminary study on mercury contamination to the environment from artisanal zinc smelting using indigenous methods in Hezhang country, Guizhou, China–Part 1: Mercury emission from zinc smelting and its influences on the surface waters. Atmospheric Environment. 2004;**38**(36):6223-6230

[70] Muller G. Index of geoaccumulation in sediments of the Rhine River. Geology Journal. 1969;**2**(3):109-118

[71] Wedepohl KH. The composition of the continental crust. Geochimica et Cosmochimica Acta. 1995;**59**(7):1217-1232

[72] Tomlinson DL, Wilson CR, Harris CR, Jeffrey DW. Problems in the assessment of heavy-metal levels in the estuaries and the formation of a pollution index. Hergoland Marine Research. 1980;**33**(1-4):566-575

[73] Karbassi AR, Bayati I, Moatta F. Origin and chemical partitioning of heavy metals in riverbed sediments. International Journal of Environmental Science and Technology. 2006;**3**(1):35-42

[74] Priju CP, Narayana AC. Heavy and trace metals in Vembanad lake sediments. International Journal of Environmental Research. 2007;**1**(4):280-289

[75] Rabee AM, Al-Fatlawy YF, Najim AA, Nameer M. Using pollution load index (PLI) and geoaccumulation index (I–geo) for the assessment of heavy metals pollution in Tigris river sediment in Baghdad region. Journal of Al-Nahrain University. 2011;**14**(4):108-114

[76] Harikumar PS, Jisha TS. Distribution pattern of trace metal pollutants in the sediments of an urban wetland in the southwest coast of India. International Journal of Engineering Science and Technology. 2019;**2**(5):840-850

[77] Chakravarty M, Patgiri AD. Metal pollution assessment in sediments of the Dikrong River, NE India. Journal of Human Ecology. 2009;**27**(1):63-67

[78] Seshan BRR, Natesan U, Deepthi K. Geochemical and statistical approach for evaluation of heavy metal pollution in core sediments in southeast coast of India. International Journal of Environmental Science and Technology. 2010;**7**(2):291-306

[79] Long ER, MacDonald DD, Smith SL, Calder FD. Incidence of adverse

biological effects within range of chemical concentrations in marine and estuarine sediments. Environmental Management. 1995;**19**(1):18-97

[80] Smith SL, MacDonald DD, Keenleyside KA, Ingersoll CG, Field J. A preliminary evaluation of sediment quality assessment values for freshwater ecosystems. Journal of Great Lakes Research. 1996;**22**(3):624-638

[81] Schaffner M, Bader HP, Scheidegger R. Modeling the contribution of point sources and non-point sources to Thachin River water pollution. Science of the Total Environment. 2009;**407**(17):4902-4915

[82] O'Regan PR. The use of contemporary information technologies for coastal research and management—A review. Journal of Coastal Research. 1996;**12**(1):192-204

[83] Zhou F, Guo HC, Hao ZJ. Spatial distribution of heavy metals in Hong Kong's marine sediments and their human impacts: A GIS-based chemometric approach. Marine Pollution Bulletin. 2007;**54**(9):1372-1384

[84] Davis HT, Aelion CM, McDermott S, Lawson AB. Identifying natural and anthropogenic sources of metals in urban and rural soils using GIS-based data, PCA, and spatial interpolation. Environmental Pollution. 2009;**157**(8-9):2378-2385

[85] Zhou F, Guo HC, Liu L. Quantitative identification and source apportionment of anthropogenic heavy metals in marine sediment of Hong Kong. Environmental Geology. 2007;**53**(2):295-305

[86] Poggio L, Borut V, Schulin R, Hepperle E, Marsan FA. Metals pollution and human bioaccessibility of topsoils in Grugliasco (Italy). Environmental Pollution. 2009;**157**(2):680-689

[87] Langran G. A review of temporal database research and its use in GIS applications. International Journal of Geographical Information System. 1989;**3**(3):215-232

[88] Carrara A, Cardinali M, Detti R, Guzzeti F, Pasqui V, Reichenbach P. GIS techniques and statistical models in evaluating landslide hazard. Earth Surface Processess Landforms. 1991;**16**(5):427-445

[89] Facchinelli A, Sacchi E, Mallen L. Multivariate statistical and GIS-based approach to identify heavy metal sources in soils. Environmental Pollution. 2001;**114**(3):313-324

[90] Yin K, Lin ZF, Ke ZY. Temporal and spatial distribution of dissolved oxygen in the Pearl River Estuary and adjacent coastal waters. Continental Shelf Research. 2004;**24**(16):1935-1948

[91] Lee CSL, Li XD, Shi WZ, Cheung SCN, Thornton I. Metal contamination in urban, suburban and country park soils of Hong Kong: A study based on GIS and multivariate statistics. The Science of the Total Environment. 2006;**356**(1-3):45-61

[92] Chowdary VM, Rao NH, Sarma PBS. Decision support framework for assessment of non-point-source pollution of groundwater in large irrigation projects. Agricultural Water Management. 2005;**75**(3):194-225

[93] Labonne M, Othman DB, Luck JM. Pb isotopes in mussels as tracers of metal sources and water movements in a lagoon (Thau Basin, S. France). Chemical Geology. 2001;**181**(1):181-191

[94] Goodwin TH, Young AR, Holmes MGR, Old GH, Hewitt N, Leeks GJL, et al. The temporal and spatial variability of sediment transport and yields within the Bradford Beck catchment, West Yorkshire. Science of the Total Environment. 2003;**314-316**:475-494

[95] Gelinas Y, Schmit JP. Extending the use of the stable lead isotope ratios as a tracer in bioavailability studies. Environmental Science and Technology. 1997;**31**(7):1968-1972

[96] Ip CCM, Li XD, Zhang G, Wong CSC, Zhang WL. Heavy metal and Pb isotopic compositions of aquatic organisms in the Pearl River Estuary, South China. Environmental Pollution. 2005;**138**(3):495-505

[97] Ong MC, Kamaruzaman MI, Yong JC, Kamaruzzaman BY, Joseph B. Metals contamination using *Polymesoda expansa* (Marsh Clam) as bio–indicator in Kelantan River, Malaysia. Malaysian Journal of Analytical Sciences. 2017;**21**(3):597-604

[98] Ong MC, Amalina I. Determination of selected metallic element in marsh clam, *Polymesoda expansa*, collected from Tanjung Lumpur mangrove forest, Kuantan, Pahang. Borneo Journal of Marine Science & Aquaculture. 2017;**1**(1):65-70

[99] Phillips DJH. Arsenic in aquatic organisms: A review emphasising chemical speciation. Aquatic Toxicology. 1990;**16**(3):151-186

[100] Rainbow PS. Trace metal accumulation in marine invertebrates: Marine biology or marine chemistry? Journal of the Marine Biological Association of the United Kingdom. 1997;77:195-210

[101] Ong MC, Kamaruzaman MI, Siti Noorhidayah A, Joseph B. Trace metals in highly commercial fishes caught along coastal water of Setiu, Terengganu, Malaysia. International Journal of Applied Chemistry. 2016;**12**(4):773-784

[102] Ong MC, Gan SL. Assessment of metallic trace elements in the muscles and fins of four landed elasmobranchs from Kuala Terengganu waters, Malaysia. Marine Pollution Bulletin. 2017;**124**(2):1001-1005

[103] Blais JM, Macdonald RW, Mackay D, Webster E, Harvey C, Smol JP. Biological mediated transport of contaminants to aquatic systems. Environmental Science and Technology. 2007;**41**(4):1075-1084

[104] Erickson RJ, Nichols JW, Cook PM, Ankley GT. Chapter 2. Bioavailability of chemical contaminants in aquatic systems. The Toxicology of Fishes. 2008:9-54

[105] Liu JK, He X. Quatitative and qualitative aspects of fish corp in relation to environmental quality. Ecotoxicology and Environmental Safety. 1987;**13**(1):61-75

[106] Anderson RS. Lack of hemocyte chemiluminescence stimulation by Perkinsus marinus in eastern oysters Crassostrea irginica with dermo disease. Journal of Aquatic Animal Health. 1999;**11**(2):179-182

[107] Bryan GW, Langston WJ, Hummerstone LG, Burt GR. A Guide to the Assessment of Heavy Metal Contamination in Estuaries Using Biological Indicators. Plymouth: Occasional Publication of the Marine Biological Association 4, Marine Biology of the United Kingdom; 1985. 92 p

[108] Rainbow PS. Biomonitoring of heavy metal availability in the marine environment. Marine Pollution Bulletin. 1995;**31**(4-12):183-192

[109] Bryan GW, Langston WJ, Hummerstone LG. The use of biological indicators of heavy metal contamination in estuaries: With special reference to an assessment of the biological availability of metals in estuarine sediments from south-west Britain. Marine Biological Association of the United Kingdom. 1980;**1**:73

[110] Butler PA, Andren L, Bonde GJ, Jernelov A, Reisch DJ. Monitoring organisms. In: Ruivo M, editor. Food and Agricultural Organisation Technical Conference on Marine Pollution and its Effects on Living Resources and Fishing, Rome, 1970. Supplement 1: Methods of Detection, Measurement and Monitoring of Pollutants in the Marine Environment. London: Fishing News Books; 1971. pp. 101-112

Chapter 5

Nitrogen and Phosphorus Eutrophication in Marine Ecosystems

Lucy Ngatia, Johnny M. Grace III, Daniel Moriasi and Robert Taylor

Abstract

Nitrogen (N) and phosphorus (P) eutrophication in marine ecosystems is a global problem. Marine eutrophication has a negative impact on food security, ecosystem health and economy through disruptions in tourism, fisheries and health industries. Both N and P have known point and non-point sources. Control of point sources has been easier than non-point sources particularly agricultural sources for both N and P as well as fossil fuel combustion for N, which remains a major challenge. Implementing mitigation strategies for N has been reported to be effective for P mitigation; however, the converse is not true due to mobility and volatility of N. Excessive N and P cause algae blooms, anoxic conditions, and ocean acidification with these conditions leading to dead zones, fish kill, toxin production, altered plant species diversity, food web disruption, tourism disruption and health issues. Management of N and P pollution includes reduction of leaching from farms through crop selection, timely and precise application of fertilizer and building artificial wetlands, proper management of animal waste, reduction of fossil fuel N emission, mitigating N and P from urban sources and restoration of aquatic ecosystem. Mitigation measures need to focus on dual nutrient strategy for successful N and P reduction.

Keywords: agriculture, eutrophication, marine, mitigation, nitrogen, phosphorus, pollution

1. Introduction

In the past few decades there have been massive increase in marine eutrophication globally [1]. The major drivers of marine eutrophication are nitrogen (N) and phosphorus (P) [2]. Eutrophication leads to hypoxia and anoxia, reduced water quality, alteration of food web structure, habitat degradation, loss of biodiversity and noxious and harmful algal blooms [1, 3]. In addition, coastal hypoxia contributes to ocean acidification harming the calcifying organisms for example mollusks and crustaceans [4].

Nitrogen and P are required to support aquatic plant growth and have been reported as the key limiting nutrients in most aquatic ecosystems. Further, N is needed for protein synthesis while as P is required for DNA, RNA and energy transfers [5]. Marine ecosystem heavily loaded with nutrients can display N limitation, P limitation and co-limitation [6] the limiting nutrient could change both seasonally and spatially [7].

A number of factors make N more limiting in the marine ecosystem than in fresh water ecosystem with two primary factors being (1) desorption of P bound to clay as salinity increase and (2) reduced/lack of planktonic N fixation as a result of increased salinity, resulting in flux of relatively P rich N poor marine water [4].

Increased N and P fertilizer and manure application in agricultural production have significantly improved crop yields and food security for the increasing human population, however fertilizer application on farms has led to serious problems with aquatic eutrophication (**Figure 1**) [1, 9]. As a result N and P in fertilizer and manure enter freshwater systems and are transported by streams and rivers to coastal areas resulting in eutrophication of coastal and marine ecosystems globally (**Figure 2**) [10–12]. In addition, atmospheric deposition of N from fossil fuel combustion contributes to the global budget for reactive N and is the largest single source of nitrogen pollution in some regions (**Table 1**) [1]. The chapter addresses the forms of N and P, and their sources. Consequences of eutrophication and mitigation strategies as well as some of the challenges faced during the mitigation process.

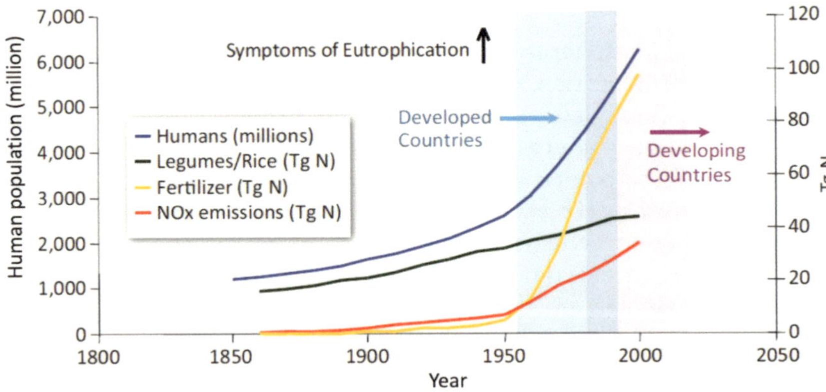

Figure 1.
Period in which the symptoms of eutrophication and hypoxia/anoxia began in developed countries and the more recent evolution of these symptoms in developing countries, modified from Schlesinger [8].

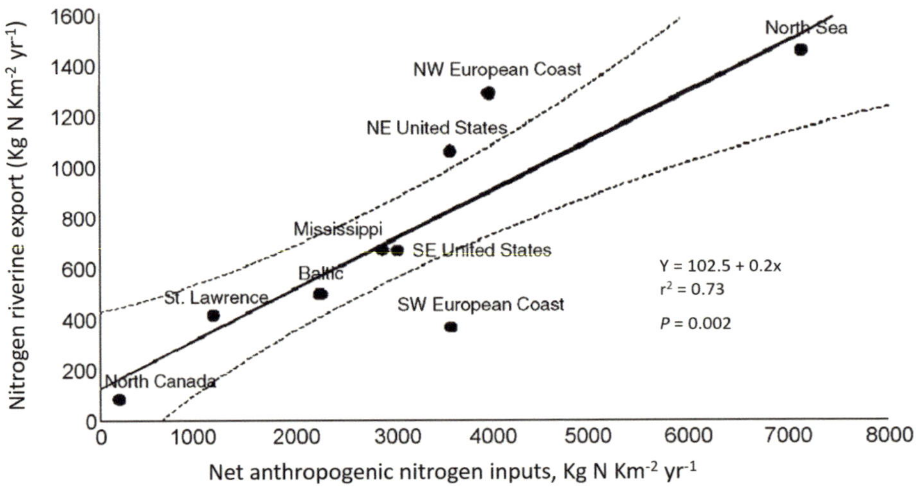

Figure 2.
Average annual nitrogen export per area of watershed from large regions around the North Atlantic Ocean to the coastal ocean as a function of net anthropogenic nitrogen inputs to the landscape per area. Modified from Howarth [1].

	1961	1997
Input		
N fixation in agricultural systems	4.9	5.9
Inorganic N fertilizer	3.1	11.2
NOx emissions from fossil fuel combustion	3.8	6.9
Total input	11.8	24.0
Exports		
Export in rivers	3.0	5.0
Atmospheric advection to oceans	0.7	1.3
Food and feed export	0.6	2.2
Total exports	4.3	8.5
Denitrification and storage	7.5	15.5
Net anthropogenic N inputs	10.5	20.5

Table 1.
Budgets indicating reactive nitrogen from human sources in United States of America (Tg N per year). Net anthropogenic N inputs indicate use of inorganic fertilizer plus N fixation in agricultural systems plus NO_x deposition from fossil fuel combustion minus the net export of N_x in food and feeds. Modified from Howarth et al. [55].

2. Forms of nitrogen and phosphorus

Most of the N on earth is molecular dinitrogen (N_2) and most of it is in the atmosphere, however, a portion of it is dissolved in the ocean [1]. Only ~0.002% of N on earth is present in living tissues and detrital organic matter [8]. Nitrogen is essential for life; however, biologically available forms such as nitrate, nitrite and ammonium are a small proportion of N on earth and as a result, N limits primary productivity in coastal marine ecosystems [13].

Soil P exists in a range of organic and inorganic compounds that differ remarkably in their biological availability in the soil environment [14]. The inorganic P compounds preferentially couple with crystalline and amorphous forms of Al, Fe, and Ca [15] the coupling is highly influenced by soil pH [16]. Organic P in most soils is dominated by a mixture of phosphate diesters (mainly nucleic acids and phospholipids) and phosphate monoesters (example; mononucleotides, inositol phosphates) with smaller amounts of phosphonates (compounds with a direct carbon–phosphorus bond) and organic polyphosphates (for example; adenosine triphosphate) [17]. Plants have the capacity to manipulate their acquisition of P from organic compounds through various mechanisms, some of which allow plants to utilize organic P as efficiently as inorganic phosphate [18]. Alkaline pH can alter the availability of P binding sites on ferric complexes as a result of competition between hydroxyl ions and bound phosphate ions [19]. Anaerobic conditions favor release of P as a result of reduction of ferric to ferrous iron [20]. While, the presence of sulfate could lead to reaction of ferric iron with sulfate and sulfide to form ferrous iron and iron sulfide leading to release of P [21]. Temperature increase can reduce adsorption of P by mineral complexes in the sediment [22]. Other physio-chemical processes affecting release of P from the sediment include pH potential, redox, reservoir hydrology and environmental conditions [23]. These physio-chemical processes could further be complicated by the influence of biological processes such as mineralization, leading to a complex system governing the release of P across sediment water interface [23].

Unlike N fertilizer, P fertilizer is not volatile, consequently very little P could be distributed from cropland to nearby terrestrial ecosystem [24]. However, excessive

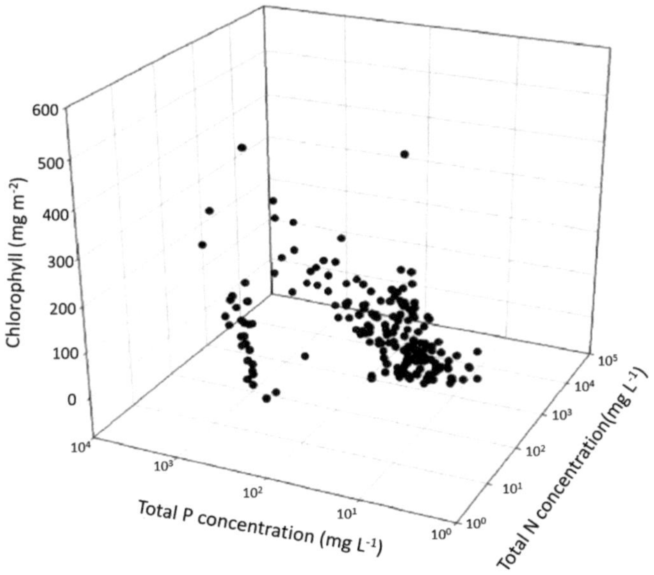

Figure 3.
Summer mean benthic chlorophyll concentrations from streams worldwide as a function of summer mean concentrations of total P and total N in the water column. Modified from Dodds and Smith [31].

P fertilizer application could result in significant transfer of P to adjacent freshwater bodies, followed by transport to coastal waters [25]. The nitrogen cycle contains diverse gaseous forms, both dissolved and particulate forms of N, while P cycle is dominated by particulate and non-gaseous forms of P [26]. This means that N pool can exchange with and escape to the atmosphere but P is trapped in receiving marine waters. The processes controlling losses of N to the atmosphere include ammonification, denitrification, nitrous and nitric oxide production and products of anaerobic ammonium oxidation (or anammox) reaction, while N fixation represents a gain from the atmosphere [27, 28]. Nitrification and denitrification are regulated by oxygen concentration and potentially can produce nitrous oxide, a climate relevant atmospheric trace gas [29]. However, there are no analogous air-water exchanges that exist in the P cycle. Therefore, while the net effect of the microbially mediated dissolved gaseous fluxes on N is loss of N to the atmosphere, P remains in the system, either as dissolved or as particulate forms [26]. While many efforts have focused on P mitigation, less attention has been given to N mitigation [30]. However, it is clear that N and P together describe eutrophication better than either can alone (**Figure 3**).

3. Sources of nitrogen and phosphorus

A century ago, the world reactive N was derived mainly through microorganisms fixation. This is the natural N fixation from the atmosphere. Currently, most of reactive N is derived from anthropogenic activities, mainly synthetic N fertilizers, manure application and fossil fuel combustion [32–36]. In addition, anthropogenic activities have accelerated biological N fixation associated with agriculture [33]. It is estimated that globally deposition of reactive N is ~25–33 Tg N per year from fossil fuel combustion, ~118 Tg N per year from fertilizer, and ~65 Tg N per year from fixation of atmospheric N_2 by cultivated leguminous crops and rice. Only ~22% of total human input on N ends up accumulating in soils and biomass, whereas ~35% enter oceans through atmospheric deposition (17%) and leaching through river runoff (18%) [25]. However, the only source of atmospheric P deposition is through mineral aerosols and the global flux is estimated at 3–4 Tg P per year [25]. Agricultural and urbanization activities are the major drivers of N pollution in the coastal waters [13]. The green revolution has led to synthetic N fertilizers, creating reactive N at a rate four times greater than fossil fuel combustion [34, 36]. Dinitrogen fixation by planktonic cyanobacteria is less likely in coastal seas compared to lakes, due to high salinity, whereby coastal planktonic N_2 fixation has not been observed at salinities higher than 8–10 and normally ocean salinity is ~35 [5].

Phosphorus sources can be natural which includes indigenous soil P, atmospheric deposition and anthropogenic P [37]. Phosphorus sources include both point and non-point sources [38]. Excess phosphorus inputs to lakes/rivers, which are eventually translocated to the marine ecosystem, usually come from industrial discharges, construction sites, urban areas, sewage and runoff from agriculture [39]. Many countries have implemented mechanisms to control point source P, however, controlling non-point P sources especially agricultural sources remains a challenge [38, 40]. The major source of nonpoint P input to water bodies is the excessive application of fertilizer or manure on farms which cause P accumulation in soils [40]. It should be noted that crop and livestock production systems are the major cause of human alteration of the global N and P cycles [41].

4. Consequences of nitrogen and phosphorus eutrophication

Eutrophication leads to excessive plant production, blooms of harmful algae, increased frequency of anoxic events, and death of fish. These conditions lead to health implications and economic losses, including losses of fish and wildlife production and losses of recreational amenities [38, 42].

4.1 Ocean acidification

Coastal hypoxia contributes to ocean acidification harming the calcifying organisms such as mollusks and crustaceans [4]. Anoxic and hypoxic water are associated with elevated carbon dioxide which causes acidification accelerating perturbation of ocean chemistry and influencing carbon dioxide emission into the atmosphere [4].

4.2 Dead zones

Hypoxia and anoxia lead to dead zones whereby fauna is eliminated or diversity and abundance is reduced. Dead zones in the coastal area have spread significantly since 1960s (**Figure 4**) and the increases are triggered by increases in primary production as a result of increased marine eutrophication fueled by riverine runoff of fertilizers and burning of fossil fuels [44]. The increased primary production lead to the accumulation of particulate organic matter which, accelerate microbial activity and consumption of dissolved oxygen in bottom waters resulting in death of fish and other marine fauna.

4.3 Human/animal health

Ranging from United States to Japan, the Black Sea and Chinese coastal waters increased nutrient loading in marine waters has been attributed to development of biomass blooms, which lead to toxic or harmful impacts on ecosystems, human health and/or recreation [45]. Approximately 60–80 species of about 400 known phytoplankton are toxin producing and capable of producing harmful algal blooms [46]. Toxin producing algae could cause mortalities of fish, birds, and marine

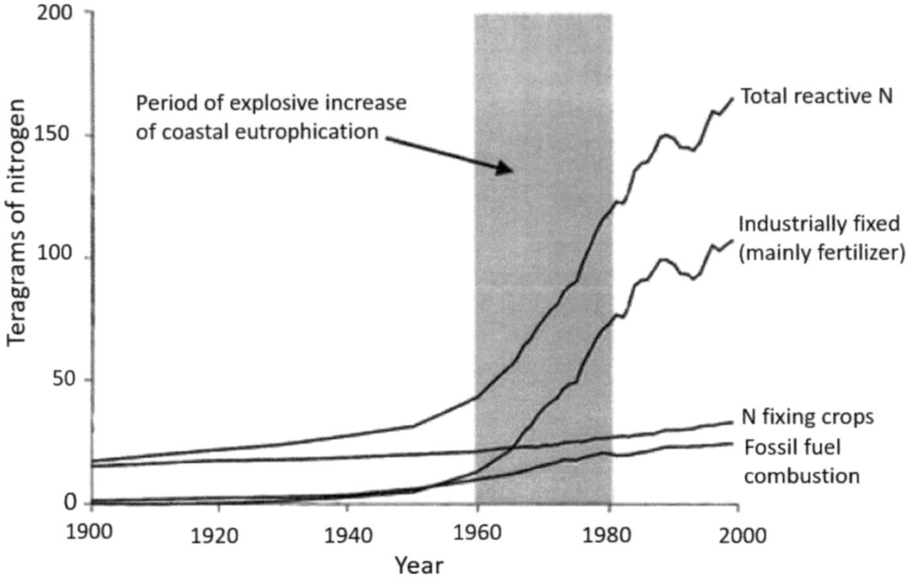

Figure 4.
Period of the explosive increase in coastal eutrophication in relation to global additions of anthropogenically fixed N from Boesch [43].

mammals as well as human illness through consumption of fisheries [47]. In humans, toxins arising from harmful algal blooms have mainly been reported from shellfish consumption [47] since bivalve shellfish (*Mollusca*) graze on algae and concentrate toxins effectively. In May and June 1998 the mortality of over 200 California sea lions (*Zalophus californianus*) and signs of neurological dysfunction in surviving sea lions along the central California coast was attributed to a harmful algal blooms [48].

4.4 Tourism

Coastal areas are hotspots for tourism, and are an important economic source for tourism [49]. The algal bloom resulting from N and P eutrophication have degraded the investment environment and damaged the tourism and hospitality industry (**Figure 3**) [50].

5. Management of nitrogen and phosphorus pollution and challenges faced

Generally, it has been much easier to manage point sources of both N and P however, non-point sources have been a challenge to control and are the main sources of pollution in the marine ecosystem [51, 52]. Nitrogen has higher mobility in the environment compared to P since N flows easily through both ground water and atmosphere [51]. It has been generally indicated that management practices for reducing N pollution in most cases are also effective in phosphorus control, however the converse is not true (**Table 2**) [51, 53]. Due to high N mobility and volatility, in some cases it might need different/additional mitigation strategies compared with P. In United States and Europe major progress has been made in reducing N pollution from municipal waste water sources which is a point source but very little progress has been made in reducing non-point source N and P pollution [51, 54]. The following are some technical solutions for N and P management.

	Effects on P reduction	Effects on N reduction
Agricultural systems		
Winter cover crops	Effective	Very effective
No till agriculture	Very effective	Not effective
Perennial cropping systems	Effective	Very effective
Buffer strips along streams	Effective	Effective only if groundwater flows are intercepted by rooting zone
Wastewater treatment		
Conventional septic system	Very effective	Not effective
Chemical precipitation advanced wastewater treatment plants	Very effective	Little effective
Denitrification advanced wastewater plants	Effective	Very effective

Table 2.
Relative effectiveness of some representative best management practices (for reducing nitrogen and phosphorus pollution of surface and groundwater. Modified from Howarth [51].

5.1 Leaching and runoff from agricultural fields

In United States, N inputs to agricultural fields doubled from 8 to 17 million metric tons per year between 1961 and 1997 [55]. Approximately 20% of the new N inputs to agricultural fields is leached to ground or surface water [55, 56]. Climate has an influence on N losses, whereby, losses are higher under high rainfall intensities and wetter years [57]. Significant amount of P from agricultural fields is lost through leachate and runoff to rivers, lakes and reservoirs and eventually finds its way to the marine ecosystem [38].

5.1.1 Growing perennial crops

Growing grasses or alfalfa rather than annuals such as soybeans or corn is presented here as a potentially beneficial practice to control N and P nonpoint source pollution. Perennials are known to maintain N in the rooting zone, thereby reducing losses to groundwater. Previous work has shown that fields planted with perennial alfalfa lost ~30–50 times less nitrate compared to fields planted with soybeans and corn in Iowa and Minnesota, USA [57, 58].

5.1.2 Planting winter cover crops

Winter cover crops provide a range of services that are beneficial to managing N and P pollution since these crops protect the soil during this vulnerable season with lower evapotranspiration rates and higher antecedent soil moisture conditions. Cover crops provide cover to soils vulnerable to accelerated erosion losses and reduce soil erosion, which is the primary mechanism whereby nutrients are transported to surface water systems. In particular, N and P are attached to soil particles that are transported via storm water runoff directly to receiving surface waters such as streams, rivers, lakes, wetlands, and oceans. Further, winter cover crops reduce nutrients transported such as leaching of nitrate into groundwater during winter and spring, this is the period that most leaching occurs in many climates [51] as a result of the antecedent moisture conditions due to reduced plant soil water and nutrient uptake. The previous literature reports that long term winter cover crops have the capability of reduced nitrate loss as much as 3-fold [59].

5.1.3 Effective N and P application

Nitrogen and P fertilizer application timing is critical to managing nonpoint source pollution in cropping systems. Fertilizer application as close as possible to the time of crop need during the growing season is presented as a viable control for pollution challenges related to N and P. In particular, fall fertilizer application is often not the most agronomically, economically, or environmentally efficient or sustainable practice since much of this amendment would be either have direct connectivity to receiving surface waters through storm water runoff during the dormant season or eventually leach to ground waters since plant uptake is limited to negligible amounts because, with the exception of fall or winter cash crops, growth begins in the spring. The previous literature has found that fall application of fertilizer can result in as much as 30–40% leaching of the fertilizer [57].

5.1.4 Optimal fertilizer application rates

Application of the right quantity of the fertilizer is critical to responsible nutrient management in most all ecosystems. Prudent nutrient management and sound economics requires that fertilizer amendment to the point of optimal yield without excess application that results in increased nutrient export via surface runoff or leaching. Incremental application rates above the point where more fertilizer application increases crop yield and N and P are no longer limiting has no effect on production [60] and can actually have negative return on yields as well as the excessive fertilizer is susceptible to surface storm water runoff and leaching to groundwater (**Figure 5**).

5.1.5 Effective buffer strips and/or forest buffers

Buffers in the form of grass hedges, stiff grass hedges, field buffers, riparian forests, forest buffers, and wetland forests are commonly applied to minimize the effects of agricultural and urban land uses. Forested conditions are optimal for control of nonpoint source pollutions due to a range of characteristics that minimize the storm runoff and subsequent soil and nutrient transport that can result [61]. These characteristics include increased infiltration rates, increased vegetative cover, a thick organic layer, increased surface roughness, and higher evapotranspiration rates among other attributes. Each of these characteristics singularly reduces the quantity of surface water available to transport nutrients as well as nutrients

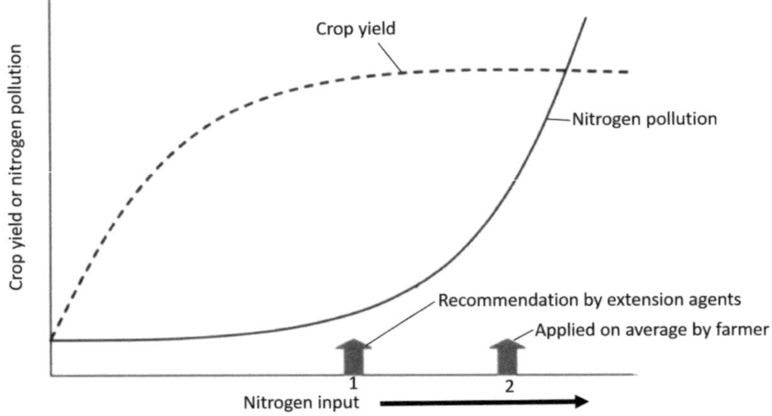

Figure 5.
Schematic representation of crop yield and reactive nitrogen export to surface and groundwater as a function of reactive nitrogen inputs to agricultural field. Arrows indicate level of fertilization recommended by extension agents (1) and actual levels applied on average by farmers (2). Modified from Howarth et al. [55].

available to leach to groundwater. Uptake of grass and/or tree species can remove excess nutrients as a means of production of a component crop whether it come in the form of biomass, wildlife habitat, timber, or hay among other economic benefits.

5.1.6 Artificial wetlands/constructed wetlands

Artificial or constructed wetlands can receive storm water runoff with increased nutrient concentrations and/or intercept tile drainage on farms by serving as a N and P sink. These wetlands reduce flux of nitrate and phosphate to receiving surface waters [62].

5.2 Animal production and concentrated animal feeding operations

Animal waste is a major contributor of N and P to coastal waters [51]. In the United States waste from animals in feedlots tends to be spread on the farm through land application, held in lagoons, or recently it is being composited [51, 53].

5.2.1 Manure application

Manure has been considered as fertilizer; however, applying manure at a rate appropriate to crop needs is a challenge as a result of uncertainty in time of nutrient release and difficulty in uniform distribution of manure (**Figure 6**) [53].

5.2.2 Lagoons hold animal waste

Lagoons can be used for holding animal waste as a retention area to prevent unacceptable pulse of nutrient loads to receiving waters. Lagoons retain nutrients until a time when they can strategically and effectively utilized, however diligence is required due to there could be significant leakage on N to groundwater and volatilization of N as ammonia to the atmosphere. The volatilized N contributes to the flux of N to marine system [53].

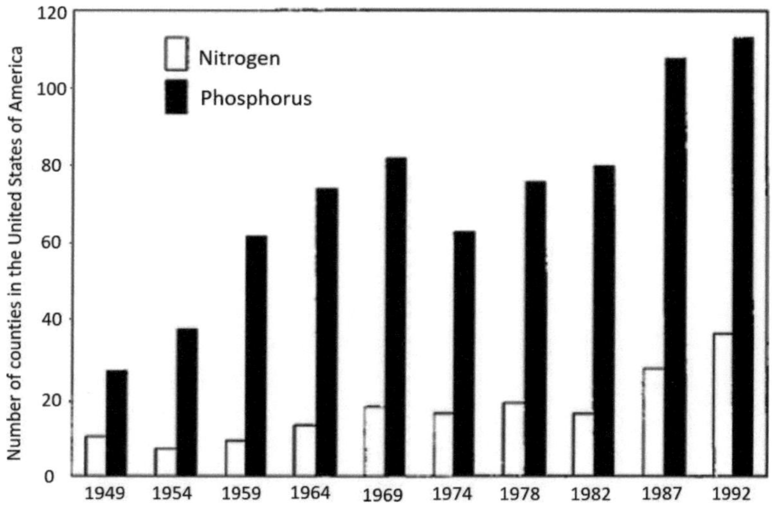

Figure 6.
Graphical representation of number of counties where manure nutrients exceed the potential plant uptake and removal, including pastureland application, modified from Howarth et al. [55].

5.2.3 Composting of animal waste

After animal waste is composited, it becomes easier to use it as effective fertilizers. However, during composting ammonia volatilization takes place, which contributes to the flux of N to marine system. In addition N volatilization lowers the quality of compost as fertilizer [53].

5.3 Fossil fuel sources of nitrogen

Fossil fuel combustion emits oxidized forms of N (NO_x) to the atmosphere [51]. In United States fossil fuel combustion emits ~6.9 million metric tons of N per year to the environment (**Figure 7**). This is ~60% of the rate of N fertilizer use in the country [55]. Most of the N emitted from fossil fuel is deposited back on the landscape in rain and dry deposition with a significant contribution to nutrient pollution in coastal water [51]. Approximately half of the fossil fuel N emission comes from mobile sources, which includes automobiles, busses, trucks and off road vehicles [51]. Electric power generation produced ~42% of reactive N [63]. Reduction of N emission from fossil fuel combustion could be achieved through; encouraging less driving and more energy efficient vehicles and remove NO_x from exhaust using catalytic converters [51]. Stricter emission standards should be applied to sports utility vehicles, trucks and off road vehicles [51]. Promote generation of electric power using fuel cells rather than traditional combustion would mitigate NO_x emissions [35]. In addition, electric power plants built for example in the United States before Clear Air Act could be modified to modern standards [63].

5.4 Urban and suburban sources

Approximately 29% of population of the United States is served by septic tanks instead of sewers [61]. In some coastal areas, septic tanks are the primary sources of N to coastal water [64]. Reduction of N leakage from septic tanks in coastal areas could be accomplished through replacing septic system with sewers and nutrient removal sewage treatment [51].

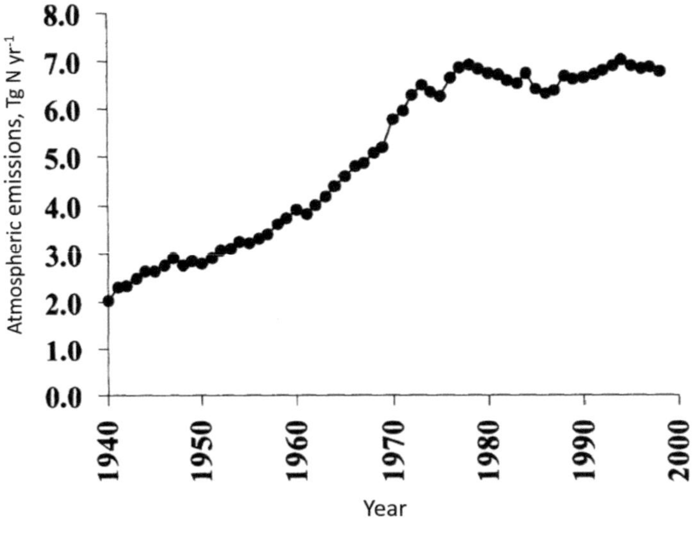

Figure 7.
Atmospheric NO_x emissions in the United States of America from 1940 to 2000, modified from Howarth et al. [55].

5.5 Aquatic ecosystem restoration

5.5.1 Wetlands as nitrogen and phosphorus interceptors: enhancing sinks

Wetlands, riparian zones and ponds can act as N and P traps, they can sediment out particulate N and P and convert reactive N into harmless N_2 through the process of denitrification [51–53]. Wetlands play a key role in P removal as a result of processes that include uptake by microbes and vegetation, peat accretion (sorption and burial in soil and sediments), and precipitation by iron and aluminum [65–67].

5.5.2 Riparian area restoration

Both N and P have been reported to contaminate groundwater [68], but s there is no clear evidence of the location and the source of the contaminant. Therefore, measures to decrease groundwater and surface water derived P and N call for the need to implement measures in the riparian area eliminate groundwater N and P directly before it enters the water bodies, this will avoid translocation of the pollutants to the marine ecosystem.

5.5.3 River/lake maintenance and restoration

River and lake rehabilitation and restoration improves their capacity to retain nutrients [69] and buffer nutrient fluxes to reduce N and P concentration in surface water reaching marine ecosystems.

6. Conclusion

Eutrophication is a major problem in the marine ecosystems that is driven primarily by N and P loading. Nitrogen and P pollution are contributed from both point and non-point sources. Many countries have been successful at controlling point sources for both N and P, however non-point sources remains a major problem. Phosphorus input into freshwater and subsequently marine systems are a result of anthropogenic activities that result in excess P with agricultural fertilizers serving as the leading cause of excess P. Historically, reactive N was contributed primarily by microorganisms and microbial activity. In recent years, reactive N is contributed by a myriad of anthropogenic activities which are led by fertilizer and manure application and fossil fuel inputs. While P is transported to the marine ecosystem in particulate and dissolved form, N is translocated as particulate, dissolved and gaseous form. The volatility and mobility of N makes it difficult to control and manage. In marine ecosystem, the salinity condition favors availability of P while limiting planktonic N fixation. This indicates that translocated N has greater effect on marine ecosystem than onsite fixed N. The consequence of N and P loading in freshwater and marine systems is multifaceted with eutrophication perhaps having the greatest impact. Eutrophication can result in increased incidence and significance of algae bloom, anoxic conditions, ocean acidification, and altered plant species diversity. The effects of eutrophication are production of toxins that affect human and animal health, fish kills that negatively impact the food security, food web disruption, and dead zones that disrupt ecosystem functioning. Further, the disruptions in ecosystem functions translate to negative impacts to the tourism industry with economic consequences. Therefore, eutrophication mitigation is essential in order to prevent excessive N and P loading to lotic and lentic systems from upslope anthropogenic activities and subsequently reaches marine ecosystems. The previous literature and investigations indicate that mitigation strategies should focus on both N and P loading reduction to ensure sustainability. Mitigation strategies to reduce the impact of N & P loading include reduction of leaching from agricultural activities, growing perennial plants, prudent application of fertilizers, and planting winter cover crops to reduce nutrient leaching via increased plant uptake. Strategies include policy and/or industry standard modifications that are likely more controversial yet likely the key to sustainable practices. These strategies include reduction of fossil fuel N emissions through transportation demand reductions, vehicle efficiency advancements, enhanced removal of NO_x from exhaust, and stricter emission standards for the transportation system. Further, terrestrial and aquatic ecosystem restoration in terms of flora and fauna are critical to ensure greater quantities N and P are effectively trapped or tied up prior to reaching marine ecosystems. Overall, measures to control eutrophication need to focus on dual nutrient reduction, instead of focusing on N or P alone in order to ensure sustainability, unless there is evidence that focusing on only one nutrient is justifiable for a given ecosystem.

Acknowledgements

The authors' special thanks go to College of Agriculture and Food Sciences, Florida A&M University for providing a conducive environment for writing this book chapter. This work was supported by USDA/NIFA (1890 Evans-Allen Research Program), USDA-Forest Service grant number 17-CA-11330140-027 and USDA-ARS grant number 58-3070-7-009.

Author details

Lucy Ngatia[1*], Johnny M. Grace III[2], Daniel Moriasi[3] and Robert Taylor[1]

1 College of Agriculture and Food Sciences, Florida A&M University, Tallahassee, FL, USA

2 USDA-Forest Service, Southern Research Station, Tallahassee, FL, USA

3 USDA-ARS Grazinglands Research Laboratory, El Reno, OK, USA

*Address all correspondence to: lucy.ngatia@famu.edu

IntechOpen

© 2019 The Author(s). Licensee IntechOpen. This chapter is distributed under the terms of the Creative Commons Attribution License (http://creativecommons.org/licenses/by/3.0), which permits unrestricted use, distribution, and reproduction in any medium, provided the original work is properly cited.

References

[1] Howarth RW. Coastal nitrogen pollution: A review of sources and trends globally and regionally. Harmful Algae. 2008;**8**:14-20

[2] Elser JJ, Bracken MES, Cleland EE, Gruner DS, Stanley W, Harpole WS, et al. Global analysis of nitrogen and phosphorus limitation of primary producers in freshwater, marine and terrestrial ecosystems. Ecology Letters. 2007;**10**:1135-1142

[3] Rabalais NN, Turner RE, Dıaz RJ, Justić D. Global change and eutrophication of coastal waters. ICES Journal of Marine Science. 2009;**66**:1528-1537

[4] Howarth R, Chan F, Conley DJ, Garnier J, Doney S, Marino R, et al. Coupled biogeochemical cycles: Eutrophication and hypoxia in temperate estuaries and coastal marine ecosystems. Frontiers in Ecology and the Environment. 2011;**9**(1):18-26

[5] Conley DJ, Paerl HW, Howarth RW, Boesch DF, Seitzinger SP, Havens KE, et al. Controlling eutrophication: Nitrogen and phosphorus. Science. 2009;**323**:1014-1015

[6] Conley DJ. Biogeochemical nutrient cycles and nutrient management strategies. Hydrobiologia. 2000;**410**:87-96

[7] Malone TC, Conley DJ, Fisher TR, Glibert PM, Harding LW. Scales of nutrient-limited phytoplankton productivity in Chesapeake Bay. Estuaries. 1996;**19**(2B):371-385

[8] Schlesinger WH. Biogeochemistry: An Analysis of Global Change. New York: Academic Press; 1997

[9] Huang J, Xu C, Ridoutt BG, Wang X, Ren P. Nitrogen and phosphorus losses and eutrophication potential associated with fertilizer application to cropland in China. Journal of Cleaner Production. 2017;**159**:171-179

[10] Li J, Glibert PM, Zhou MJ, Lu SH, Lu DD. Relationships between nitrogen and phosphorus forms and ratios and the development of dinoflagellate blooms in the East China Sea. Marine Ecology Progress Series. 2009;**383**:11-26

[11] Smith LED, Siciliano G. A comprehensive review of constraints to improved management of fertilizers in China and mitigation of diffuse water pollution from agriculture. Agriculture, Ecosystems and Environment. 2015;**209**:15-25

[12] Xu H, Paerl HW, Qin BQ, Zhu GW, Gao G. Nitrogen and phosphorus inputs control phytoplankton growth in eutrophic Lake Taihu, China. Limnology and Oceanography. 2010;**55**:420-432

[13] Howarth RW, Marino RM. Nitrogen as the limiting nutrient for eutrophication in coastal marine ecosystems: Evolving views over 3 decades. Limnology and Oceanography. 2006;**51**:364-376

[14] Condron LM, Turner BL, Cade-Menun BJ. The chemistry and dynamics of soil organic phosphorus. In: Sims JT, Sharpley AN, editors. Phosphorus: Agriculture and the Environment. Madison, Wisconsin, USA: ASA-CSSASSSA; 2005. pp. 87-121

[15] Negassa W, Leinweber P. How does the Hedley sequential phosphorus fractionation reflect impacts of land use and management on soil phosphorus: A review. Journal of Plant Nutrition and Soil Science. 2009;**172**:305-325

[16] Ngatia LW, Hsieh YP, Nemours D, Fu R, Taylor RW. Potential phosphorus eutrophication mitigation strategy: Biochar carbon composition,

thermal stability and pH influence phosphorus sorption. Chemosphere. 2017;**180**:201-211

[17] Turner BL. Organic phosphorus transfer from terrestrial to aquatic environments. In: Turner BL, Frossard E, Baldwin DS, editors. Organic Phosphorus in the Environment. Wallingford: CAB International; 2005. pp. 269-295

[18] Tarafdar JC, Claassen N. Organic phosphorus compounds as a phosphorus source for higher plants through the activity of phosphatases produced by plant roots and microorganisms. Biology and Fertility of Soils. 1988;**5**:308-312

[19] De Montigny C, Prairie Y. The relative importance of biological and chemical processes in the release of phosphorus from a highly organic sediment. Hydrobiologia. 1993;**253**:141-150

[20] Bostrom B. Potential mobility of phosphorus in different types of lake sediment. Internationale Revue der Gesamten Hydrobiologie. 1984;**69**:457-475

[21] Bostrom B, Andersen JM, Fleischer S, Jansson M. Exchange of phosphorus across the sediment water interface. Hydrobiologia. 1988;**170**:229-244

[22] Redshaw CJ, Mason CF, Hayes CR, Roberts RD. Factors influencing phosphate exchange across the sediment-water interface of eutrophic reservoirs. Hydrobiologia. 1990;**192**:233-245

[23] Perkins RG, Underwood GJC. The potential for phosphorus release across the sediment-water interface in a eutrophic reservoir dosed with ferric sulphate. Water Research. 2001;**35**(6):1399-1406

[24] Penuelas J, Sardans J, Rivas-Ubach A, Janssens IA. The human-induced imbalance between C, N and P in Earth's life system. Global Change Biology. 2012;**189**:5-8

[25] Penuelas J, Poulter B, Sardans J, et al. Human-induced nitrogen–phosphorus imbalances alter natural and managed ecosystems across the globe. Nature Communications. 2013, 2013;**4**:2934

[26] Paerl HW. Controlling eutrophication along the freshwater–marine continuum: Dual nutrient (N and P) reductions are essential. Estuaries and Coasts. 2009;**32**:593-601

[27] Codispoti LA, Brandes JA, Christensen JP, Devol AH, Naqvi SWA, Paerl HW, et al. The oceanic fixed nitrogen and nitrous oxide budgets: Moving targets as we enter the anthropocene? Scientia Marina. 2001;**65**(2):85-105

[28] Rich J, Dale OR, Song B, Ward BB. Anaerobic ammonium oxidation (Anammox) in Chesapeake Bay sediments. Microbial Ecology. 2008;**55**:311-320

[29] Voss M, Bange HW, Dippner JW, Middelburg JJ, Montoya JP, Ward B. The marine nitrogen cycle: Recent discoveries, uncertainties and the potential relevance of climate change. Philosophical Transactions of the Royal Society B. 2013;**368**(1621). DOI: 10.1098/rstb.2013.0121

[30] Lewis WM Jr, Wurtsbaugh WA, Paerl HW. Rationale for control of anthropogenic nitrogen and phosphorus to reduce eutrophication of inland waters. Environmental Science and Technology. 2011;**45**:10300-10305

[31] Dodds WK, Smith VH. Nitrogen, phosphorus, and eutrophication in streams. Inland Waters. 2016;**6**(2):155-164

[32] Choi E, Yun Z, Chung TH. Strong nitrogenous and agro-wastewater:

Current technological overview and future direction. Water Science and Technology. 2004;**49**(5-6):1-5

[33] Galloway JN, Dentener FJ, Capone EW, et al. Nitrogen cycles: Past, present, and future. Biogeochemistry. 2004;**70**:153-226

[34] Galloway JN, Cowling EB. Reactive nitrogen and the world: 200 years of change. Ambio. 2002;**31**:64-71

[35] Howarth RW. Nitrogen cycle. In: Encyclopedia of Global Environmental Change. Chichester, U.K: Wiley; 2001

[36] Vitousek PM, Aber JD, Howarth RW, Likens GE, Matson PA, Schindler DW, et al. Human alteration of the global nitrogen cycle: Sources and consequences. Ecological Applications. 1997;**7**:737-750

[37] Haygarth PM, Condron LM, Heathwaite AL, Turner BL, Harris GP. The phosphorus transfer continuum: Linking source to impact with an interdisciplinary and multi-scaled approach. Science of the Total Environment. 2005;**344**:5-14

[38] Carpenter SR, Caraco NF, Correll DL, Howarth RW, Sharpley AN, Smith VH. Nonpoint pollution of surface waters with phosphorus and nitrogen. Ecological Applications. 1998;**8**:559-568

[39] Carpenter SR. Eutrophication of aquatic ecosystems: Bistability and soil phosphorus. Proceedings of the National Academy of Sciences of the United States of America. 2005;**102**:10002-10005

[40] Bennett EM, Carpenter SR, Caraco NF. Human impact on erodable phosphorus and eutrophication: A global perspective: Increasing accumulation of phosphorus in soil threatens rivers, lakes, and coastal oceans with eutrophication. Bioscience. 2001;**51**(3):227-234

[41] Bouwman L, Goldewijk KK, Van Der Hoek KW, Beusen AHW, Van Vuurena DP, Willems J, et al. Exploring global changes in nitrogen and phosphorus cycles in agriculture induced by livestock production over the 1900-2050 period. Proceedings of the National Academy of Sciences of the United States of America. 2013;**110**(52):20882-20887

[42] Wilson MA, Carpenter SR. Economic valuation of freshwater ecosystem services in the United States: 1971-1997. Ecological Applications. 1999;**9**:772-783

[43] Boesch DF. Challenges and opportunities for science in reducing nutrient over enrichment of coastal ecosystems. Estuaries. 2002;**25**(4b):886-900

[44] Diaz RJ, Rosenberg R. Spreading dead zones and consequences for marine ecosystems. Science. 2008;**321**:926-929

[45] Anderson DM, Glibert PM, Burkholder JM. Harmful algal blooms and eutrophication: Nutrient sources, composition and consequences. Estuaries. 2002;**25**(4b):704-726

[46] Smayda TJ. Harmful algal blooms: Their ecophysiology and general relevance to phytoplankton blooms in the sea. Limnology Oceanography. 1997;**42**:1137-1153

[47] James KJ, Carey B, O'Halloran JO, Van Pelt FNAM, Skrabakova Z. Shellfish toxicity: Human health implications of marine algal toxins. Epidemiology and Infection. 2010;**138**:927-940

[48] Scholin CA et al. Mortality of sea lions along the Central California coast linked to a toxic diatom bloom. Nature. 2000;**403**:80-84

[49] Diercks-Horn S, Metfies K, Jackel S, Medlin LK. The ALGADEC device: A semi-automated rRNA biosensor for the

detection of toxic algae. Harmful Algae. 2011;**10**:395-401

[50] Le C, Zha Y, Li Y, Sun D, Lu H, Yin B. Eutrophication of Lake waters in China: Cost, causes, and control. Environmental Management. 2010;**45**:662-668

[51] Howarth RW. The development of policy approaches for reducing nitrogen pollution to coastal waters of the USA. Science in China. Series C, Life Sciences. 2005;**48**:791-806

[52] Howarth RW, Walker D, Sharpley A. Sources of nitrogen pollution to coastal waters of the United States. Estuaries. 2002;**25**:656-676

[53] NRC. Clean Coastal Waters: Understanding and Reducing the Effects of Nutrient Pollution. Washington: National Academy Press; 2000

[54] Boesch DF, Brinsfield RB, Magnien RE. Chesapeake Bay eutrophication: Scientific understanding, ecosystem restoration, and challenges for agriculture. Journal of Environmental Quality. 2001;**30**:303-320

[55] Howarth RW, Boyer EW, Pabich WJ, Galloway JN. Nitrogen use in the United States from 1961 to 2000 and potential future trends. Ambio. 2002;**31**:88-96

[56] Howarth RW, Billen G, Swaney D. Regional nitrogen budgets and riverine N & P fluxes for the drainages to the North Atlantic Ocean: Natural and human influences. Biogeochemistry. 1996;**35**:75-139

[57] Randall GW, Mulla DJ. Nitrate nitrogen in surface waters as influence by climatic conditions and agricultural practices. Journal of Environmental Quality. 2001;**30**:337-344

[58] Randall GW, Huggins DR, Russelle MP, et al. Nitrate losses through subsurface tile drainage in CRP, alfalfa, and row crop systems. Journal of Environmental Quality. 1997;**26**:1240-1247

[59] Staver KW, Brinsfield RB. Use of cereal grain winter cover crops to reduce groundwater nitrate contamination in the Mid-Atlantic coastal plain. Journal of Soil and Water Conservation. 1998;**53**:230-240

[60] NRC. Managing Wastewater in Coastal Urban Areas. Washington, DC: National Academy Press; 1993

[61] Grace JM. Forest operations and water quality in the south. Transactions of ASAE. 2005;**48**(2):871-880

[62] Mitsch WJ, Day J, Gilliam JW, Groffman PM, et al. Reducing nitrogen loading to the Gulf of Mexico from the Mississippi River basin: Strategies to counter a persistent ecological problem. Bioscience. 2001;**51**:373-388

[63] Ryther JH, Dunstan WM. Nitrogen, phosphorus, and eutrophication in the coastal marine environment. Science. 1971;**171**:1008-1012

[64] Valiela I, Geist M, McClelland J, et al. Nitrogen loading frm watersheds to estuaries: Verification of the Waquoit Bay nitrogen loading model. Biogeochemistry. 2000;**49**:277-293

[65] Kadlec RH, Wallace S. Treatment Wetlands. 2nd ed. Boca Raton, Florida: CRC Press; 2008

[66] Mitsch WJ, Gosselink JG. Wetlands. 4th ed. New York: John Wiley & Sons, Inc.; 2011

[67] Vymazal J. Removal of nutrients in various types of constructed wetlands. Science of the Total Environment. 2007;**380**:48-65

[68] McMahon A, Santos IR. Nitrogen enrichment and speciation in

a coral reef lagoon driven by groundwater inputs of bird guano. Journal of Geophysical Research. 2017;**122**(9):7218-7236

[69] Schoumans OF, Chardon WJ, Bechmann ME, Gascuel-Odoux C, Hofman G, Kronvang B, et al. Mitigation options to reduce phosphorus losses from the agricultural sector and improve surface water quality: A review. Science of the Total Environment. 2014;**468-469**:1255-1266

Chapter 6

Decadal Pollution Assessment and Monitoring along the Kenya Coast

Eric Ochieng Okuku, Kiteresi Linet Imbayi, Owato Gilbert Omondi, Wanjeri Veronica Ogolla Wayayi, Mwalugha Catherine Sezi, Kombo Mokeira Maureen, Stephen Mwangi and Nancy Oduor

Abstract

Marine contamination arising from land-based sources is on the rise along the Kenyan Coast. We carried out a decadal pollution survey between 2008 and 2018 to determine the levels of various pollutants (nutrients, trace metals, persistent organic pollutants, and ^{210}Po) in water, sediment, and biota collected from selected locations in Kenya. Nutrient levels in water ranged between <0.10 and 1560.00, <0.10 and 1320.00, and <0.10 and 3280.00 μg/L for PO_4^{3-}-P, (NO_2^- + NO_3^-)-N, and NH_4^+-N, respectively, while Chl-a values ranged between 0.02 and 119.37 mg/L. Total PAH, PCBs, and OCPs in sediment from the studied locations ranged from BDL-37800, 0.012–7.99 and BDL-6.10 ng/g. High level of PAH in Kilindini port was primarily from petroleum sources. DDD + DDE/DDT ratio was above 0.5 suggesting historical input. Sediment trace metal concentration from selected locations in Kenyan estuaries had various ranges, that is, Al (0.06–9804284.00 μg/g), Zn (3.82–367.20 μg/g), Cu (7.5–169.60), Cd (DL –2.40 μg/g), Mn (BDL-169.60 μg/g), Cr (2.55–239.10 μg/g), and Pb (BDL-135.60) μg/g dw. Surface sediment ^{210}Po activities ranged between 20.29 and 43.44 Bq kg^{-1} dw. Chl-a and PO_4^{3-}-P data revealed enhance primary productivity in Mombasa peri-urban creeks and estuarine areas. Although the reported concentrations of trace metals and POPs are low in most locations from Kenya, there is a potential risk of bioaccumulation of these contaminants in marine biota; thus, there is a need for continuous monitoring to protect both ecosystem and human health.

Keywords: marine pollution, nutrients, heavy metal

1. Introduction

Kenya is experiencing an increase in human population, urbanization, and industrialization which is characterized by massive utilization of natural resources coupled with an increase in waste production [1]. UNEP GEF-funded WIOLaB project identified that over 70% of pollutants entering marine and coastal ecosystems originate from the land [2]. Many pollutants from these land-based sources (such as sewage, oil hydrocarbons, sediments, nutrients, pesticides, litter, and plastics and toxic wastes) enter the sea via rivers and surface runoff [3]. For instance, sewage is still being indiscriminately discharged into peri-urban creeks of Mombasa [4].

Kenya's coastal waters are thus continuously enriched with large amounts of nutrients, particularly nitrogen (N) and phosphorus (P), silica (Si) and sulfur (S) from anthropogenic activities and wastewater from households and industries [4] despite the fact that effluent pollution had been identified as the most serious of all land-based threats to the marine environment and an area which requires more attention [5]. Excessive nutrient enrichment results in nutrient pollution and eutrophication with severe economic, environmental, and ecological implications to the marine environment due to the shift in the Redfield ratio [6]. Some of the impacts of nutrient enrichment include increased abundance and biomass of phytoplankton and macroalgae [7]; reduced light penetration; increased development of hypoxic, anoxic, and dead zones [8, 9]; changes in phytoplankton community structure and alteration of food chains and webs [10]; development of harmful algal blooms that may produce substances toxic to aquatic organisms or humans [11]; dieback of seagrasses, algal beds, and corals [12, 13]; loss of biodiversity; and increased incidences and duration of harmful algal blooms [14].

The other class of contaminants that are discharged through sewage, surface runoff, and industrial discharge includes heavy metals and persistent organic pollutants (POPs). Trace metals such as lead (Pb), cadmium (Cd), copper (Cu), chromium (Cr), nickel (Ni), mercury (Hg), arsenic (As), and zinc (Zn) are not easily degraded in the natural environment and are bioaccumulated along the food chain [15]. They are also known to affect productivity, reproduction, and survival of marine organisms and can be hazardous to human health at elevated concentrations [15].

Persistent contamination of organic pollutants (POPs), such as organochlorine pesticides (OCPs), polycyclic aromatic hydrocarbons (PAHs), and polychlorinated biphenyls (PCBs) in the aquatic environment is a worldwide problem [16]. The anthropogenic input of these POPs in the Indian Ocean includes industrial and sewage effluents, storm water drains, shipping activities, spillage, rivers, atmospheric fallout, coastal activities, and natural oil seeps. For instance, these compounds have been used extensively in Kenya for various applications in industries and agriculture [17]. They are characterized by long-term stability (not easily degradable in the environment and can persist in sediments for decades or even centuries), bioaccumulative nature, long-range transport capability, and may have high toxic effects on aquatic living organisms [18, 19]. These chemicals have since been replaced with organophosphate and carbamate, which are less persistent in the environment under the Stockholm Convention on Persistent Organic Pollutants [20, 21]. However, their residues are still present in the marine environment.

Among the natural radionuclides, alpha emitters (e.g. ^{210}Po) have been reported to have significant radiological effects resulting from their accumulation in organisms [22]. ^{210}Po is a naturally occurring radioactive material (NORM) alpha-emitter occurring in very low concentrations in the environment as a part of the uranium decay chain. ^{210}Po has a radioactive half-life of 138.4 days and is produced in the marine environment from the decay of ^{210}Pb (which is a daughter isotope derived from ^{226}Ra dissolved in seawater). ^{210}Pb can also be introduced into marine environment directly from the atmosphere from the decay of ^{222}Rn [22, 23]. It is ubiquitously distributed in rocks, soils, making up earth's crust, in the atmosphere and in natural waters [24]. It can also be derived in insignificant quantities from lead-containing wastes from uranium, vanadium, and radium refining operations [25]. The activity of ^{210}Po in the environment has increased in the past due to human activity such as fossil fuel combustion, use of phosphate fertilizers in agriculture, and discharge of domestic and industrial sewage.

UNEP-GEF WIO-LaB Project identified municipal and industrial effluents, contaminated surface runoff including groundwater and agricultural runoff [2] as the major sources of land-based pollution and microbial contamination, suspended solids, chemical pollution, marine litter (including debris), and eutrophication (harmful/nuisance algal blooms) as priority pollution categories in the WIO region. In Kenya, limited action has been taken to control or manage pollution. For instance, only a couple of studies have been carried out to determine long-term pollution status or their impacts on marine and coastal ecosystems and the potential human health.

During the assessment and monitoring program, water matrix was analyzed for nutrient pollution whereas sediments were analyzed for trace metals, radionuclides, and POP contamination. Sediment was the preferred matrix for persistent pollutants given that it a known sink and an archive of pollutants brought into the aquatic environment from direct discharges, surface runoff, atmospheric fallout, and nonpoint source [19, 26, 27]. The distribution of these contaminants in sediments can serve as a useful index of pollution and potential environmental risks [17, 28].

2. Materials and methods

Water and sediment samples were collected from the estuarine areas of Ramisi-Vanga system in the Kenyan South Coast; peri-urban creeks of Mombasa; and estuarine areas of the North Coast (**Figure 1**). Ramisi-Vanga system (Ramisi, Mwena, and Umba) is a low-lying coastal plain submergent complex (below 30 m contour) with extensive cover of mangrove forest, intertidal areas covered with seagrass beds and shallow water lagoons harboring the coral reefs. These critical systems are interlinked through the exchange of water, nutrients, and carbon by the tidally controlled circulation and river discharge [29]. The Kisite-Mpunguti Marine Protected Area (KMMPA) and Shimoni are situated approximately 90 km south of Mombasa with very few scattered fringe mangrove cover on the shoreline though there is no river input into this system apart from sewage and runoff [30]. The area constitutes an important tourist attraction and important resource for the surrounding communities [31]. The main characterizing features of interest around Mombasa peri-urban creeks (Makupa, Kilindini, Mtwapa, and Tudor creeks) include the Kilindini port and the Mombasa Municipal refuse dumpsite, (Kibarani), both of which are found in the vicinity of the Makupa creek stations and the Kenya Meat Commission (KMC) beef factory located within the vicinity of the Tudor creek [32]. Mtwapa creek is situated 25 km north of Mombasa. It is a tidal creek lined by mangrove forests and extensive mud banks. The creek receives freshwater input through seasonal rivers and it is reported to be relatively eutrophic due to direct release of raw sewage into the creek at the vicinity of Shimo la Tewa government prison [30]. These peri-urban creeks were chosen due to anthropogenic activities in the adjacent areas while riverine estuarine systems were chosen given that river-derived material like water, sediments, and pollutants have a tremendous ecological and toxicological influence on coastal zones [35]. North Coast sites comprised of estuarine areas of River Tana which arises from Aberdare range and Mount Kenya passing through the arid and semiarid areas of the eastern part of the country and finally drain into the Indian Ocean in a fan-shaped delta [34] while Sabaki river originates from Athi river where it joins Tsavo river and forms river Galana where it also discharges into the sea. These two rivers affect the marine ecological conditions of Watamu, Malindi, and Lamu [33].

Nutrient sample collection, preservation, shipment, storage, and analysis were done following procedures described by Parsons [36], APHA [37] and measured using Genesys 10S Vis spectroscopy (Thermo Scientific™). Sediment samples were collected using Uwitec corer (fitted with a 50-cm plastic sampler, Ø 8 cm) following the procedures provided in IAEA TECDOC-1415. The samples were transported to the laboratory awaiting analysis.

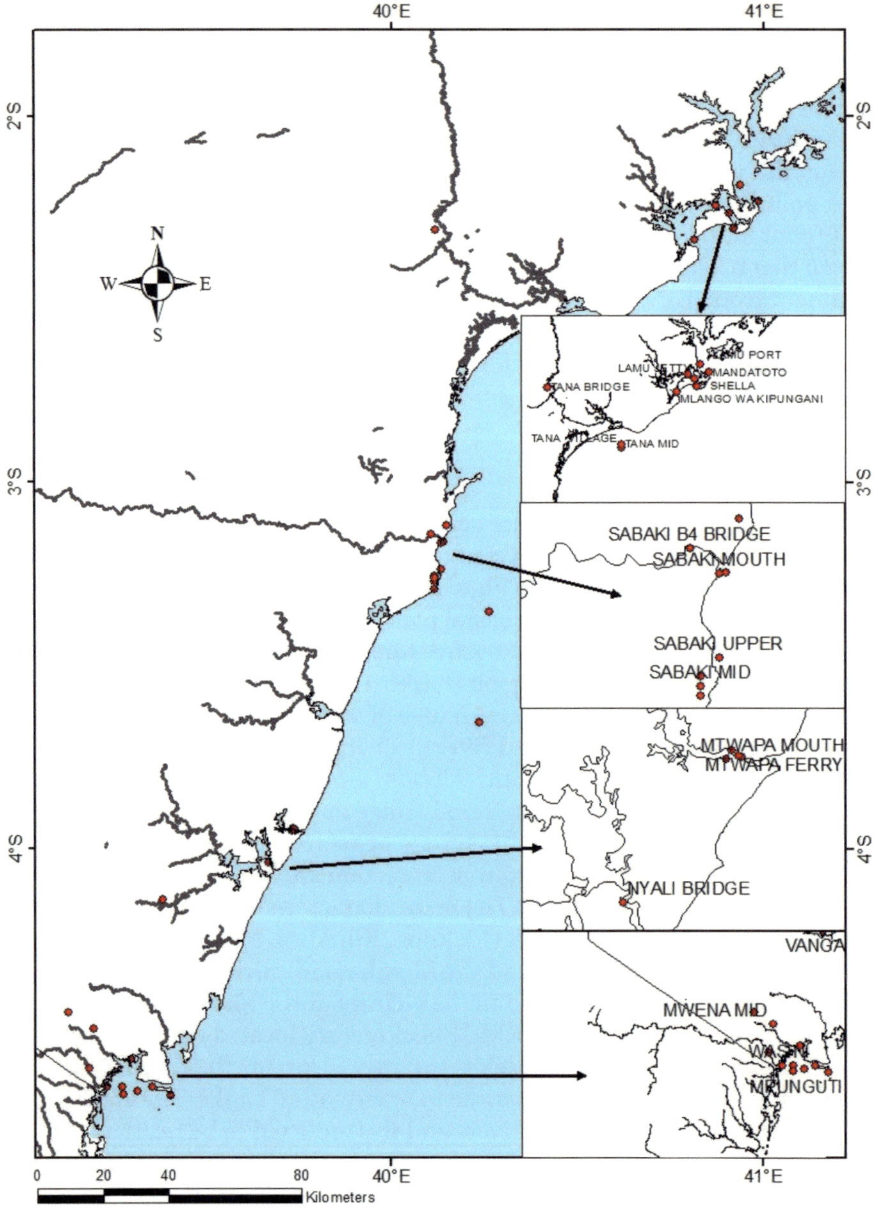

Figure 1.
Map showing sampling sites: Estuaries (Umba, Ramisi, Sabaki, Mwena, and Tana); north coast (Kilifi, Malindi, and Lamu); south coast (Kibuyuni, Shimoni, Wasini, Sii Island, Gazi, and Kisite); and peri-urban creeks of Mombasa (Mtwapa, Tudor, and Makupa).

For analysis of trace metals and alpha-emitter ^{210}Po radionuclide, top 5 cm of sediment for each site were pooled, mixed by hand, and freeze-dried, homogenized sieved, dry-weighed, and then digested in a microwave (Microwave Accelerated Reaction system). Elemental analysis of Cu, Zn, Mn, Pb, Cd, and Cr was carried out using inductively coupled plasma mass spectrometer (ICP-MS) (Varian, Australia) while ^{210}Po activity was measured using alpha-spectrometry with silicon surface barrier detectors (EG and G) coupled to a PC running Maestro TM data acquisition software [38].

PCBs, OCPs, and PAH were analyzed as described by Thompson et al. [39] where wet samples were freeze-dried using Heto powerdry LL3000 freeze dryer (Thermo

Scientific), then known weight of the sediment sample was spiked with internal standard for PCBs (PCB 30, PCB 103, PCB 155, PCB 198 from Promochem, Dr. Ehrenstorfer GmbH, France), OCPs (d8 4,4′ DDT from Cambridge Isotope Laboratory, France), and PAH (per-deuterated PAHs; Phenanthrene-d10, benzo [a]pyrene- d12, benzo [e] pyrene d12 and benzo [g, h, i]perylene-d12 from Cambridge Isotope Laboratories, Andover, USA; and Fluoranthene-d10, chrysene-d12 and pyrene-d10 from MSD isotopes, Division of Merck Frost Canada INC, Montreal, CND) to quantify the recoveries. The samples were then extracted using START E microwave-assisted extraction system (Milestone, Italy). The extract was concentrated using Rapidvap LABCONCO (Serlabo Technologies, France) and then concentrated samples were subjected to the clean-up process. Thereafter, the extracts were finally concentrated under nitrogen, transferred to 100-μl container and analyzed using GC-MS.

3. Pollutants' source identification and risk assessment

To determine the possible pollution sources of DDT, the approach by Kilunga et al. and Mohammed et al. [18, 40] (that classified sediments with (DDD +DDE)/DDT ratio greater than 0.5 as historic and ratio less than 0.5 as recent input) was applied. The ratio of DDE to DDD was used in this study to give an idea of the degradation pathway of DDTs in the sediment [41]. DDE/DDD ratio greater than 1 was indicative of anaerobic degradation of DDT into DDD while the ratio less than 1 implies DDT conversion into DDE via aerobic degradation [42].

The pollution data obtained compared with effects range-low (ER-L) and effects range-median (ER-M) values (sediment quality guidelines, SQGs) provided by MacDonald et al. and MacDonald and Ingersoll [43, 44] were used to evaluate the ecotoxicological significance/potential impacts of trace metals, PAH, PCBs, and DDT on benthic organism. ER-L represents the value at which toxicity may be observed in sensitive marine species (i.e., pollution is at a level that may not cause harm to aquatic environment) and ER-M represents the concentration below which adverse effects are expected to occur.

4. Results and discussion

4.1 Nutrient pollution

Results of this study showed high values of nutrient concentration in stations within estuaries compared to oceanic water stations (**Figure 2a–d**). The highest value of phosphate concentration was recorded in Kisite (304.82 μg/L) followed by River Sabaki (205.79 μg/L) and the least was in Kilifi (7.52 μg/L) whereas the highest values of ammonium were, however, recorded in Mombasa stations of Tudor creek (81.71 μg/L). The stations in Mombasa were also observed to have high nutrient levels compared to Malindi and Lamu (**Figure 2d** and **b**), a condition that can be attributed to the high population associated with municipal and industrial wastes in Mombasa compared to the other towns. It was observed that the nutrient levels for stations in Tana river estuary, which was hypothetically thought to be receiving higher nutrient input from the agricultural and residential population in the river catchment, were strikingly low (**Figure 2a**). This observation of low nutrient and chlorophyll-a concentration can be explained by the presence of several dams, which form the Seven Folks Dams Scheme. These dams are known to settle down suspended materials with adsorbed nutrients [45], therefore reducing the transportation of these sediments and nutrients downstream. Nutrient ranges recorded

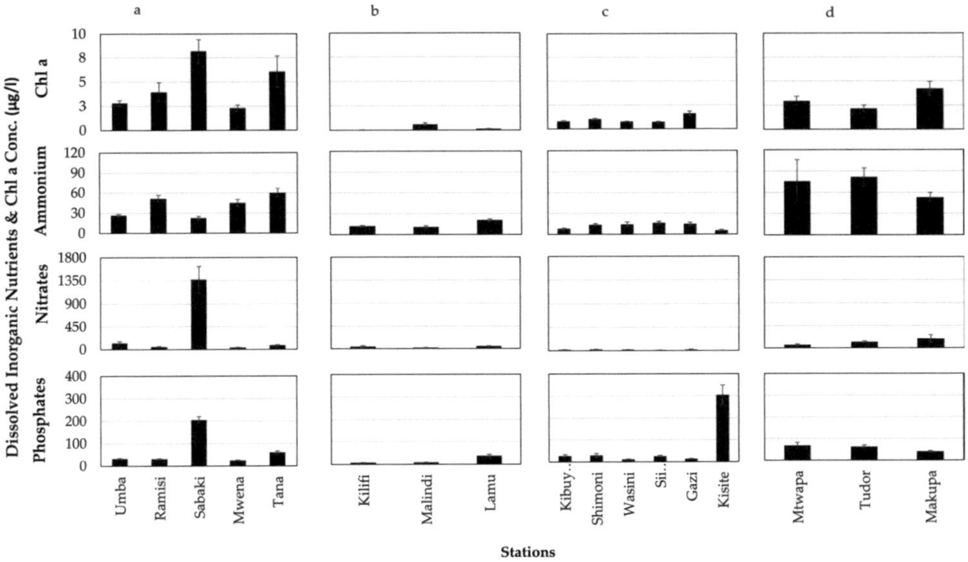

Figure 2.
Showing nutrient phosphates (µg/L), nitrates (µg/L), ammonia (µg/L), and chlorophyll-a (µg/L) distribution pattern observed in: (a) south coast estuaries: Umba, Ramisi, Sabaki, Mwena, and Tana; (b) north coast open water, stations: Kilifi, Malindi, and Lamu; (c) south coast shoreline area: Kibuy (Kibuyuni), Shimoni, Wasini, Sii Island, Gazi, and Kisite; and (d) peri-urban stations of Mombasa: Mtwapa, Tudor, and Makupa.

in this study are similar to those realized by [46, 47] in the Indian Mangroves of 14–600 and 1.00–80 µg/L for nitrates and phosphates respectively. Chlorophyll-a showed a distribution pattern almost similar to that of nutrients with low values recorded in Kilifi (0.03 µg/L) while the highest ranges were observed in River Sabaki (8.19 µg/L).

4.2 Spatial distribution of heavy metals

The aluminum (Al) levels in the peri-urban creeks had a higher concentration range: 0.06–17483.66 µg/g) compared to estuaries (range: 719484.00–9804284.00 µg/g) and pristine area Gazi Bay (335.00 µg/g). The metal concentration seemed normal since aluminum concentrations are not likely to be significantly affected by anthropogenic aluminum sources [47].

Cr concentration levels in estuaries were higher (range: 101.10–239.10 µg/g) compared to Gazi Bay (1.58 µg/g), Lamu port (range: 2.55–7.58 µg/g), and peri-urban creeks (range: 11.53–24.43 µg/g) with an exception of Kilindini harbor that had Cr concentration of 42,338.00 µg/g. The relatively high concentration of Cr in the river could have been caused by mining activities upstream whereas the high level observed in one of the stations in Kilindini could have resulted from the cement bulk storage at Kilindini harbor which is adjacent to the station [48].

Pb concentration in estuaries was higher (range: 45.70–135.60 µg/g) compared to the peri-urban creeks (range: 2.75–40.65 µg/g). These high concentrations could be attributed to both natural and anthropogenic Pb point-source discharges, primarily from its use as a fuel additive, which has made it a pervasive and persistent pollutant worldwide [49]. The concentration reported in this study was, however, lower compared to Pattani Bay (range: 79.00–97.00 µg/g), Thailand Laptev, Russia (range: 16.00–22.00 µg/g; Nolting in press) and the levels previously reported for the Kenyan Coastal Zone (range: 0.13–0.56 µg/g; by [50].

The concentration of Zn was higher in the peri-urban creeks (range: 23.36–264.30 μg/g) compared to the estuaries (range: 144.20–367.20 μg/g) and Gazi Bay (3.82 μg/g). High levels of Zn are associated to the dissolution of sacrificial zinc anodes used in marinas, ships, and on leisure boats at the Kenya ports and water sports areas; the adjacent Kibarani dumpsite is also a contributor since most types of wastes in the dumpsite find their way into Makupa creek [51].

Mn level in estuaries had higher concentrations (range: 3809.00–6781.00 μg/g) compared to peri-urban creeks (range: 16.10–58.50 μg/g) and Gazi Bay (0.77 μg/g). The same trend was observed in Cu levels in which estuaries had the highest concentration (range: 63.30–169.60 μg/g) compared to peri-urban creeks (range: (0.51–52.16 μg/g) and Gazi Bay which was BDL. Cd levels in peri-urban creeks were higher (range: BDL-2.40 μg/g), compared to the pristine area Gazi Bay (0.03 μg/g) (**Table 1**). These values may be attributed to historical industrial activity around the estuaries, that is, gypsum mining in Tana River County; five cement manufacturing companies located in Athi River: Athi River Mining, Bamburi Portland Cement, East African Portland Cement, Savanna Cement, and Simba Cement; and small landfill along Sabaki river [52, 53].

The mean ^{210}Po activities determined in the surface sediment from the three peri-urban creeks ranged between 20.29 and 43.44 Bq kg^{-1} dw. Mtwapa creek had the highest mean (37.56 ± 2.14 Bq kg^{-1} dw) compared to Makupa creek (30.42 ± 2.16 Bq kg^{-1} dw) and Tudor creek (28.64 ± 2.86 Bq kg^{-1} dw). These results are comparable to those of surface sediments from the Venice lagoon ecosystem in Italy in the range of 26–45 Bq kg^{-1} [54].

The high levels of ^{210}Po activity in Mtwapa creek could have originated from the seasonal river (River Mto Mkuu) which flows into the creek discharging finer sediment from the mainland agriculture activities. In addition, high levels of silt and organic matter were found in Mtwapa creek and this may have resulted in high levels of ^{210}Po activity since ^{210}Po tends to adsorb on a finer particle of silt (<63 μm) due to its specific surface area compared to coarse particles (>63 μm) [49, 50]. The same observation was made by Aközcan and Uğur Görgün [57] in Izmir Bay and Didim who attributed high levels of ^{210}Po to Buyuk Menderes river.

4.3 Organic pollutant pollution

PAH concentration in the surface sediment from Ramisi, Sabaki, and Tana rivers was in the range of BDL-0.16, 0.06–7.98, and 0.78–13.20 ng/g, respectively. Higher concentration of PAH in Tana river could be as a result of rapid urbanization, industrialization, gasoline or diesel fuel from fishing boats, and high erosion activities in their catchment areas. The concentrations were, however, lower compared to those reported by Chen and coworkers [58] who recorded concentration ranges of 362.00–15, 667.00 ng/g (dw) for Weihe river, a river draining arid and semiarid regions as Tana river.

The surveyed sites in Lamu, namely, Wange Fish Port, proposed Lamu Port, and Mokowe Fish port, had PAH concentration of less than 10.00 ng/g (dw). This is because the port is still under construction and un-operational. The PAH levels were lower than the levels observed in Kilindini port (range: <50.00–37,800.00 ng/g dw) with a mean concentration of 6086.00 ng/g dw. The PAH level in Kilindini was, however, higher than levels reported by Saleem et al. [59, 53] in Karachi harbor (mean: 27.52 ng/g dw) and [54, 60] in Kaohsiung harbor, Taiwan (range: 472.00–16,201.00 ng/g; mean 5764.00 ng/g). High level of PAH was observed in Kilindini due leakage of combusted and un-combusted fuel from intense shipping activities, oil spills experienced in Makupa creeks, and oil terminal activities in Kilindini

	Pristine environment	Estuaries		Peri-urban			SQG	
Sampling site	Gazi Bay	Sabaki mouth	Tana mouth	Lamu port	Makupa creek	Kilindini harbor	ER-L	ERM-
Al	335.00	9804284.00	719484.00	659.50–1674.90	17483.66	0.06–2.85	—	NA
As	—	—	—	BDL-3.54	—	<0.01–2.99	33	NG
Cd	0.027	—	—	BDL-0.13	2.4	<0.02–0.11	5	9.6
Co	0.072	—	—	—	—	—	—	—
Cr	1.576	101.10	239.10	2.55–7.58		11.53–42,338	80	370.00
Cu	BDL	63.30	169.60	0.51–2.98	52.16	6.57–15.3	70	270.00
Hg	4.623	—	—	BDL	—	<0.01	0.15	—
Mn	0.757	3809.00	6781.00	16.10–58.5	90.47	—	—	NG
Ni	—	4.90	13.10	—	—	8.47–16.22	30	—
Pb	—	45.70	135.600	BDL	40.65	2.75–11.46	35	218.00
Zn	3.82	144.20	367.20	—	264.3	23.36–38.45	120	410.00

NG: no guideline.
BDL: below detection limit.

Table 1.
Decadal comparison of trace metals of the Kenyan coastal sediment with the set international standards for sediment quality.

port as well as ferry services. This port serves many shipping vessels involved in cargo importation for Kenya as well as the landlocked countries in East Africa. The total OCPs namely HCN, Heptachlor, Hepoxide, 2,4′ DDE, Cis chlordane, Trans Nonachlor, 4,4'DDE, 2,4' DDD, 4,4' DDD, 2,4′ DDT, 4,4′ DDT, and Mirex in sediments from Ramisi, Sabaki, and Tana estuarine areas ranged between ND-0.13, 0.37–1.81, and 0.03–3.94 ng/g respectively. The concentration of OCP reported by this study (range: 0.03–3.94 ng/g) was lower compared to the levels reported for Tana estuary by Lalah et al. [61, 55] (range: <0.003–108.51 ng/g). This confirms the reduction in utilization of the products containing OCPs in Kenya.

The DDT and its metabolites 2,4′ DDE, 4,4′ DDE, 2,4' DDD, 4,4' DDD, 2,4′ DDT, and 4,4′ DDT were detected in the surface sediment in both Sabaki and Tana river estuaries. The highest concentrations of ∑6 DDTs were recorded in Tana river (11.57 ng/g dw) compared to Sabaki (2.34 ng/g dw) and Ramisi (0.23 ng/g dw). This suggests that even though DDT was banned in Kenya in 1986, there could still be illegal use of DDT in Kenya [56].

The DDE-to-DDD ratio for Ramisi, Sabaki, and Tana river estuaries were 0.20, 0.16, and 0.17 respectively, suggesting an aerobic degradation of DDT to DDD in the areas via dehydrochlorination and oxidation processes [18, 62]. The study revealed high ratios (above 0.5) for Ramisi (3.38), Sabaki (2.82), and Tana rivers (2.10), indicating a historical input of DDT into the estuaries. The historical input of DDT into marine environment has similarly been reported by Aly Salem et al. [63] for Egyptian Mediterranean Coast, as well as Lu et al. [28] for Poyang Lake, China. This shows the reduced application of DDT in some part of the world.

High concentration of PCBs was observed in River Tana estuary (range: 4.94–7.99 ng/g dw) compared to concentration in Ramisi (ND-0.15 ng/g dw) and Sabaki (0.58–2.40 ng/g dw) which could be attributed to leakage or inadvertent disposal of materials used in transformers and capacitors along the Tana river channel since Tana river holds the Seven Folks hydropower dams which generate approximately 49% of Kenyan electricity.

4.4 Ecological risk assessment of pollution along the Kenyan Coast

According to [64], some of the estuaries and peri-urban stations (Mwena, Ramisi, Sabaki, Umba, Lamu, Kibuyuni, Sii Island, and Makupa) could be classified as eutrophic given that they have phosphates levels >0.021 mg/L whereas Shimoni and Gazi could be classified as oligotrophic. Trace metal concentrations in estuaries (River Sabaki and Tana mouth) except for Pb and Cr (**Table 1**) in the sediments were below the effect range-low (ER-L)/effect range-median (ER-M). This indicates that high concentration may pose adverse effects [65].

Sediment quality criteria and concentration ranges of PAHs, PCBs, and DDT contaminants are summarized in (**Table 2**). The ∑PAHs, ∑PCBs, and ∑DDT concentration in sediment from Ramisi, Sabaki, and Kilindini were below ER-L and ER-M suggesting the low ecological risk of these compounds to sediment-dwelling

	Ramisi	Sabaki	Tana	Kilindini	Lamu	ER-L	ER-M
∑PAH	0.01–0.16	0.79–3.75	0.78–13.2	<50–37,800	<10	4000	35,000
∑PCB	0.12–0.15	0.58–2.40	4.99–7.99	<10	—	50	400
∑DDT	0.01–0.128	0.22–1.57	0.25–5.23	—	—	3	350

Table 2.
Comparison of PAH, DDT, and PCB (ng/g dw) concentrations in sediments of the study area with the sediment quality guidelines (SQG).

organisms. However, the concentration level of ∑DDT was higher than ER-L in Tana estuary but was significantly lower than the ER-M values suggesting that the accumulated DDT in sediments in Tana estuary could have a potential ecotoxicological impact on the benthic fauna.

5. Conclusion

From the study, it is evident that human influences like urbanization and agricultural activities are the major nutrient contributors to the marine environment through continuous discharge of untreated sewage effluents and agricultural runoffs delivered by rivers. Trace metal concentrations in estuaries, peri-urban creeks, and Lamu port were below the ER-L and ER-M indicating that there are no potential adverse effects on marine biota with an exception of Kilindini harbor that had a higher Cr concentration that may pose adverse effects. The concentrations of PAH, PCB, and OCP were generally high in Tana estuary as compared to the other estuaries in this study; however, the concentrations were lower than ER-L suggesting no potential impacts on organisms. It was also observed that the concentration level of ∑DDT was higher in Tana estuary than ER-L and lower for ER-M values suggesting that the accumulated persistent organic pollutants in sediments from Tana estuary could have a potential ecotoxicological impact on the benthic fauna. It is therefore important to put in place effective sewage and wastewater management strategies and implement good agricultural practices to manage the emerging pollution problem along the Kenyan Coast. A monitoring program for these pollutants is also important so as to always have an early warning.

Acknowledgements

Funding for this work was provided by Kenya Marine and Fisheries Research Institute, KMFRI) and RAF 7008, 7009, 7015 Project and CRP-K41016 projects jointly funded by International Atomic Energy Agency, IAEA and Government of Kenya. We are greatly indebted to the Directors of these Institutions for supporting this work. We also appreciate the efforts of KMFRI staff that assisted in field samples collection and analysis in one way or the other.

Author details

Eric Ochieng Okuku*, Kiteresi Linet Imbayi, Owato Gilbert Omondi, Wanjeri Veronica Ogolla Wayayi, Mwalugha Catherine Sezi, Kombo Mokeira Maureen, Stephen Mwangi and Nancy Oduor
Kenya Marine and Fisheries Research Institute, Mombasa, Kenya

*Address all correspondence to: ochiengokuku2003@yahoo.com

IntechOpen

© 2019 The Author(s). Licensee IntechOpen. This chapter is distributed under the terms of the Creative Commons Attribution License (http://creativecommons.org/licenses/by/3.0), which permits unrestricted use, distribution, and reproduction in any medium, provided the original work is properly cited.

References

[1] Okuku EO, Mubiana VK, Hagos KG, Peter HK, Blust R. Bioavailability of sediment-bound heavy metals on the East African Coast. West Indian Ocean Journal of Marine Science. 2010;**9**(1):31-42

[2] UNEP-GEF. Project Document for the UNEP-GEF WIO-LaB Project entitled 'Addressing Land-based Activities in the Western Indian Ocean. 2004

[3] Daoji L, Daler D. Ocean pollution from land-based sources: East China Sea, China. AMBIO: A Journal of the Human Environment. 2004;**33**(1):107-113. DOI: 10.1579/0044-7447-33.1.107

[4] Ohowa B, Munga D, Kiteresi LI, Wanjeri VO, Okumu S, Kilonzo J, et al. Sewage pollution in the Coastal waters of Mombasa City, Kenya: A norm rather than an exception. Int J Environ Res. 2011;(4)

[5] Centre UWCM, Group C of ML on S (Programme). DAW. Seamounts, deep-sea corals and fisheries: Vulnerability of deep-sea corals to fishing on seamounts beyond areas of national jurisdiction. UNEP/Earthprint; 2006

[6] Redfield AC. The biological control of chemical factors in the environment. American Scientist. 1958;**46**(3):230A-2221A

[7] van der Struijk LF, Kroeze C. Future trends in nutrient export to the coastal waters of South America: Implications for occurrence of eutrophication. Global Biogeochem Cycles. 2010;**24**(4). DOI: 10.1029/2009GB003572

[8] Diaz RJ, Rosenberg R. Spreading dead zones and consequences for marine ecosystems. Science (80-). 2008;**321**(5891):926-929

[9] Selman M, Greenhalgh S, Diaz R, Sugg Z. Eutrophication and hypoxia in coastal areas: A global assessment of the state of knowledge. World Resour Inst. 2008;**284**:1-6

[10] Howarth RW. In: Visgilio GR, Whitelaw DM, editors. Atmospheric Deposition and Nitrogen Pollution in Coastal Marine Ecosystems BT-Acid in the Environment: Lessons Learned and Future Prospects. Boston, MA: Springer US; 2007. pp. 97-116. DOI: 10.1007/978-0-387-37562-5_6

[11] SAB EPA. Reactive Nitrogen in the United States: An Analysis of Inputs, Flows, Consequences, and Management Options. EPA-SAB-11-013. Washington, DC: Science Advisory Board, US Environmental Protection Agency; 2011. Available from: http://yosemite.epa.gov/sab/sabproduct.nsf/WebBOARD/INCSupplemental

[12] Burkholder JM, Mason KM, Glasgow HB Jr. Water-column nitrate enrichment promotes decline of eelgrass Zostera marina: Evidence from seasonal mesocosm experiments. Marine Ecology Progress Series. 1992;**81**(2):163-178

[13] Gappa JJL, Tablado NHM. Influence of sewage pollution on a rocky intertidal community dominated by the mytilid Brachidontes. Marine Ecology Progress Series. 1990;**63**:163-175

[14] Howarth RW, Marino R, Swaney DP, Boyer EW. Wastewater and watershed influences on primary productivity and oxygen dynamics in the lower Hudson River estuary. The Hudson River Estuary. 2006;**136**

[15] Okuku EO. Bioavailability of Sediment-Bound Heavy Metals along the Coast of Eastern Africa and Evaluation of Metal Concentrations along the Trophic Hierarchy. 2007

[16] Alonso-Hernandez CM, Mesa-Albernas M, Tolosa I. Organochlorine

pesticides (OCPs) and polychlorinated biphenyls (PCBs) in sediments from the Gulf of Batabanó, Cuba. Chemosphere. 2014;**94**:36-41. Available from: http://www.sciencedirect.com/science/article/pii/S0045653513012368

[17] Barakat AO, Mostafa A, Wade TL, Sweet ST, El Sayed NB. Distribution and ecological risk of organochlorine pesticides and polychlorinated biphenyls in sediments from the Mediterranean coastal environment of Egypt. Chemosphere. 2013;**93**(3):545-554. Available from: http://www.sciencedirect.com/science/article/pii/S0045653513009065

[18] Kilunga PI, Sivalingam P, Laffite A, Grandjean D, Mulaji CK, de Alencastro LF, et al. Accumulation of toxic metals and organic micro-pollutants in sediments from tropical urban rivers, Kinshasa, Democratic Republic of the Congo. Chemosphere. 2017;**179**:37-48. Available from: http://www.sciencedirect.com/science/article/pii/S0045653517304538

[19] Miglioranza KSB, Gonzalez M, Ondarza PM, Shimabukuro VM, Isla FI, Fillmann G, et al. Assessment of Argentinean Patagonia pollution: PBDEs, OCPs and PCBs in different matrices from the Río Negro basin. Sci Total Environ. 2013;**452-453**:275-285. Available from: http://www.sciencedirect.com/science/article/pii/S0048969713002386

[20] Barasa MW, Lalah JO, Wandiga SO. Seasonal variability of persistent organochlorine pesticide residues in marine fish along the Indian Ocean coast of Kenya. Toxicol Environ Chem. 2008;**90**(3):535-547. DOI: 10.1080/02772240701556312

[21] Sun H, Qi Y, Zhang D, Li QX, Wang J. Concentrations, distribution, sources and risk assessment of organohalogenated contaminants in soils from Kenya, Eastern Africa. Environ Pollut. 2016;**209**:177-185. Available from: http://www.sciencedirect.com/science/article/pii/S0269749115302001

[22] Skwarzec B, Jahnz A. The inflow of polonium ^{210}Po from Vistula river catchments area. J Environ Sci Heal - Part A Toxic/Hazardous Subst Environ Eng. 2007;**42**(14):2117-2122

[23] Thi Van T, Duc Tam N, Duc Nhan D, Quang Long N, Van Thang D, Duy Cam B. Water-sediment distribution and behaviour of Polonium (^{210}Po) in a shallow coastal area with high concentration of dissolved organic matters in water, North Vietnam. Nucl Sci Technol. 2016;**6**(2):1-14

[24] Matthews KM, Kim C-K, Martin P. Determination of ^{210}Po in environmental materials: A review of analytical methodology. Appl Radiat Isot. 2007;**65**(3):267-279. Available from: http://www.sciencedirect.com/science/article/pii/S0969804306003290

[25] Hains A. Lead Poisoning: Triggers and Thresholds. Buffalo: NY Partnersh Public Good; 2017

[26] Arrebola JP, Cuellar M, Claure E, Quevedo M, Antelo SR, Mutch E, et al. Concentrations of organochlorine pesticides and polychlorinated biphenyls in human serum and adipose tissue from Bolivia. Environ Res. 2012;**112**:40-47. Available from: http://www.sciencedirect.com/science/article/pii/S0013935111002507

[27] Commendatore MG, Franco MA, Gomes Costa P, Castro IB, Fillmann G, Bigatti G, et al. Butyltins, polyaromatic hydrocarbons, organochlorine pesticides, and polychlorinated biphenyls in sediments and bivalve mollusks in a mid-latitude environment from the Patagonian coastal zone. Environmental Toxicology and Chemistry. 2015;**34**(12):2750-2763. DOI: 10.1002/etc.3134

[28] Lu M, Zeng D-C, Liao Y, Tong B. Distribution and characterization of organochlorine pesticides and polycyclic aromatic hydrocarbons in surface sediment from Poyang Lake, China. Sci Total Environ. 2012;**433**:491-497. Available from: http://www.sciencedirect.com/science/article/pii/S0048969712009412

[29] Kiteresi LI, Okuku EO, Mwangi SN, Ohowa B, Wanjeri VO, Okumu S, et al. The influence of land based activities on the phytoplankton communities of Shimoni-Vanga system, Kenya. International Journal of Environmental Research. 2012;**6**(1):151-162

[30] Nyunja JA, Mavuti KM, Wakwabi EO. Trophic ecology of *Sardinella gibbosa* (Pisces: Clupeidae) and *Atherinomorous lacunosus* (Pisces: Atherinidae) in Mtwapa Creek and Wasini Channel, Kenya. Western Indian Ocean Journal of Marine Science. 2002;**1**(2):181-189. Available at: http://www.oceandocs.net/handle/1834/40

[31] Meyler S, Felix H, Crouthers R. Abundance and distribution of Indo-Pacific humpback dolphins (*Sousa chinensis*) in the Shimoni Archipelago, Kenya. Western Indian Ocean Journal of Marine Science. 2011;**10**(2):201-209. Available at: http://41.215.122.106/dspace/bitstream/0/5262/1/Meyler2012.pdf

[32] Mwashote BM. Cadmium and lead levels in Mombasa, Kenya levels of cadmium and lead in water, sediments and selected fish species in Mombasa, Kenya. Western Indian Ocean Journal of Marine Science. 2003;**2**(1):25-34

[33] Hoorweg JC, Versleijen N, Wangila B, Degen A. Income diversification and fishing practices among artisanal fishers on the Malindi-Kilifi coast. 2006

[34] Okuku EO, Bouillon S, Ochiewo JO, Munyi F, Kiteresi LI, Tole M. The impacts of hydropower development on rural livelihood sustenance. International Journal of Water Resources Development. 2016;**32**(2):267-285

[35] Wu Y, Wang X, Ya M, Li Y, Hong H. Distributions of organochlorine compounds in sediments from Jiulong River Estuary and adjacent Western Taiwan Strait: Implications of transport, sources and inventories. Environ Pollut. 2016;**219**:519-527. Available from: http://www.sciencedirect.com/science/article/pii/S0269749116304699

[36] Parsons TR. A Manual of Chemical & Biological Methods for Seawater Analysis. Elsevier; 2013

[37] APHA. Standard Methods for the Examination of Water and Wastewater. 9th ed. Washington, DC: American Public Health Association, American Water Works Association and Water Environmental Federation; 1998

[38] Benedik L, Vrecek P. Determination of Pd-210 and Po-210 in environmental samples. Acta Chimica Slovenica. 2001;**48**(2):199-213

[39] Thompson S, Budzinski H, LeMenach K, Letellier M, Garrigues P. Multi-residue analysis of polycyclic aromatic hydrocarbons, polychlorobiphenyls, and organochlorine pesticides in marine sediments. Analytical and Bioanalytical Chemistry. 2002;**372**(1):196-204. DOI: 10.1007/s00216-001-1170-1

[40] Mohammed A, Peterman P, Echols K, Feltz K, Tegerdine G, Manoo A, et al. Polychlorinated biphenyls (PCBs) and organochlorine pesticides (OCPs) in harbor sediments from Sea Lots, Port-of-Spain, Trinidad and Tobago. Marine Pollution Bulletin. 2011;**62**(6):1324-1332. Available from: http://www.sciencedirect.com/science/article/pii/S0025326X1100186X

[41] Wu Q, Leung JYS, Yuan X, Huang X, Li H, Huang Z, et al. Biological

risk, source and pollution history of organochlorine pesticides (OCPs) in the sediment in Nansha mangrove, South China. Marine Pollution Bulletin. 2015;**96**(1):57-64. Available from: http://www.sciencedirect.com/science/article/pii/S0025326X15003276

[42] Ali U, Li J, Zhang G, Mahmood A, Jones KC, Malik RN. Presence, deposition flux and mass burden of persistent organic pollutants (POPs) from Mehmood Booti Drain sediments, Lahore. Ecotoxicology and Environmental Safety. 2016;**125**:9-15. Available from: http://www.sciencedirect.com/science/article/pii/S0147651315301664

[43] MacDonald DD, Ingersoll CG, Smorong DE, Lindskoog R. Development and Evaluation of Numerical Sediment Quality Assessment Guidelines for Florida Inland Waters. 2003

[44] MacDonald DD, Ingersoll CG. A guidance manual to support the assessment of contaminated sediments in freshwater ecosystems. US Environmental Protection Agency, Great Lakes National Program Office; 2002

[45] Rosenberg DM, McCully P, Pringle CM. Global-scale environmental effects of hydrological alterations: introduction. AIBS Bulletin. 2000;**50**(9):746-751

[46] Bala Krishna Prasad M. Nutrient stoichiometry and eutrophication in Indian mangroves. Environmental Earth Sciences. 2012;**67**(1):293-299. DOI: 10.1007/s12665-011-1508-8

[47] Zhang W, Feng H, Chang J, Qu J, Xie H, Yu L. Heavy metal contamination in surface sediments of Yangtze River intertidal zone: An assessment from different indexes. Environmental Pollution. 2009;**157**(5):1533-1543. Available from: http://www.sciencedirect.com/science/article/pii/S0269749109000219

[48] Kamau JN, Kuschk P, Machiwa J, Macia A, Mothes S, Mwangi S, et al. Investigating the distribution and fate of Al, Cd, Cr, Cu, Mn, Ni, Pb and Zn in sewage-impacted mangrove-fringed creeks of Kenya, Tanzania and Mozambique. Journal of Soils and Sediments. 2015;**15**(12):2453-2465. DOI: 10.1007/s11368-015-1214-3

[49] Mager EM, Esbaugh AJ, Brix KV, Ryan AC, Grosell M. Influences of water chemistry on the acute toxicity of lead to *Pimephales promelas* and *Ceriodaphnia dubia*. Comparative Biochemistry and Physiology, Part C: Toxicology & Pharmacology. 2011;**153**(1):82-90

[50] Everaarts JM, Swennen C, Cheewasedtham W. Heavy metals (Cu, Zn, Cd, Pb) in surface sediment and organisms in a short food chain from the intertidal zone of Pattani Bay, Thailand–A preliminary study. Wallaceana. 1994;**72**:17-24

[51] Obhođaš J, Valković V. Contamination of the coastal sea sediments by heavy metals. Applied Radiation and Isotopes. 2010;**68**(4):807-811. Available from: http://www.sciencedirect.com/science/article/pii/S0969804309008100

[52] Davies TC, Osano O. Sustainable mineral development: Case study from Kenya. Geological Society of London, Special Publication. 2005;**250**(1):87-93. Available from: http://sp.lyellcollection.org/content/250/1/87.abstract

[53] Johnson UK. River sediment supply, sedimentation and transport of the highly turbid sediment plume in Malindi Bay, Kenya. Journal of Geographical Sciences. 2013;**23**(3):465-489. DOI: 10.1007/s11442-013-1022-x

[54] Jia G, Belli M, Sansone U, Rosamilia S, Blasi M. ^{210}Pb and ^{210}Po concentrations in the Venice lagoon ecosystem (Italy) and the potential radiological impact to the local

public and environment. Journal of Radioanalytical and Nuclear Chemistry. 2003;**256**(3):513-528

[55] Mahmood ZUW, Mohamed CAR, Ishak AK, Bakar NSA, Ishak K. Vertical distribution of ^{210}Pb and ^{226}Ra and their activity ratio in marine sediment core of the East Malaysia coastal waters. Journal of Radioanalytical and Nuclear Chemistry. 2011;**289**(3):953-959. Available from: http://link.springer.com/10.1007/s10967-011-1206-8

[56] Venunathan N, Narayana Y. Activity of ^{210}Po and ^{210}Pb in the riverine environs of coastal Kerala on the southwest coast of India. Journal of Radiation Research and Applied Science. 2016:392-399

[57] Aközcan S, Uğur Görgün A. Trace metal and radionuclide pollution in marine sediments of the Aegean Sea (Izmir Bay and Didim). Environment and Earth Science. 1 Aug 2013;**69**(7):2351-2355

[58] Chen Y, Jia R, Yang S. Distribution and source of polycyclic aromatic hydrocarbons (PAHs) in water dissolved phase, suspended particulate matter and sediment from Weihe river in Northwest China. International Journal of Environmental Research and Public Health. 2015;**12**

[59] Saleem M, Ali Khan A, Ahmed K, Qamaruddin M, Kahkashan S, Imran Hasany S, et al. Polycyclic Aromatic Hydrocarbons (PAHs) in Surface Sediments from the Karachi Harbor and off-Clifton Coast, Karachi: Spatial Distribution, Composition and Ecological Risk Assessment. Vol. 112017. pp. 2222-3045

[60] Chen C-W, Chen C-F. Distribution, origin, and potential toxicological significance of polycyclic aromatic hydrocarbons (PAHs) in sediments of Kaohsiung Harbor, Taiwan. Marine Pollution Bulletin. 2011;**63**(5):417-423. Available from: http://www.sciencedirect.com/science/article/pii/S0025326X11002505

[61] Lalah JO, Yugi PO, Jumba IO, Wandiga SO. Organochlorine pesticide residues in Tana and Sabaki Rivers in Kenya. Bulletin of Environmental Contamination and Toxicology. 2003;**71**(2):298-307

[62] Cheng H, Lin T, Zhang G, Liu G, Zhang W, Qi S, et al. DDTs and HCHs in sediment cores from the Tibetan Plateau. Chemosphere. 2014;**94**:183-189. Available from: http://www.sciencedirect.com/science/article/pii/S0045653513013982

[63] Aly Salem DMS, Khaled A, El Nemr A. Assessment of pesticides and polychlorinated biphenyls (PCBs) in sediments of the Egyptian Mediterranean Coast. Egyptian Journal of Aquatic Research. 2013;**39**(3):141-152. Available from: http://www.sciencedirect.com/science/article/pii/S1687428513001131

[64] Siokou-Frangou I, Pagou K. Assessment of the trophic conditions and ecological status in the Inner Saronikos Gulf. InTechnical Report for the Ministry of Environment, Planning and Public Works. NCMR: Athens. Mar 2000. p. 43

[65] Benson NU, Anake WU, Essien JP, Enyong P, Olajire AA. Distribution and risk assessment of trace metals in Leptodius exarata, surface water and sediments from Douglas Creek in the Qua Iboe Estuary. Journal of Taibah University For Science. 1 May 2017;**11**(3):434-439

Chapter 7

Oil Spill Dispersion Forecasting Models

Antigoni Zafirakou

Abstract

Oil spill models are used worldwide to simulate the evolution of an oil slick that occurs after an accidental ship collision or during oil extraction or other oil tanker activities. The simulation of the transport and fate of an oil slick in the sea, by evaluating the physicochemical processes that take place between oil phase and the water column, is the base for the recognition and assessment of its environmental effects. Numerous oil spill dispersion models exist in the bibliography. The contribution of this chapter is the introduction of a 3D oil slick simulation model developed by the Aristotle University of Thessaloniki, which has been recurrently used in different updated forms and applied in operational mode, since 1991 when it was originally created. The model has been tested in various hypothetical scenarios in North Aegean Sea, Greece, and responded with great success. Findings of the present study highlight the existing experience on the subject and denote the applicability of such models in either tracing the source of a spill or in predicting its path and spread, thus proving their value in real-time crisis management.

Keywords: oil spill modeling, simulation of oil slick transport, oil physicochemical processes, operational applications, oil spill monitoring

1. Introduction

Pollution that comes from a single source, like an oil or chemical spill, is known as point source. Commonly, this type of pollution has bigger impact on the marine environment, but fortunately, it occurs less often than land runoff [1]. Oil spills occur after a collision and/or the sinking of oil tankers, under bad weather conditions, and are extremely hazardous in ports with dense maritime traffic. The environmental impact of oil accidents is immense on both the water ecosystems and the coastal environment, including the urban and economic growth of the affected coastal zones. Strong winds and significant water currents contribute to the creation of accidents and concurrently to the spreading of the spills. Ship collision or malfunction, or simply cleaning and sailing, can contribute to marine pollution by spreading garbage, black or gray water, sludge, water ballast, coatings or even air emissions, in deep sea or at seashore. The most common nautical accidents occur due to sinking or foundering, grounding, structural failure, scuttling, by contact or collision, explosion or fire, or after disappearance or abandonment [2]. The discharge from oil pipelines, oil platforms, and vessels is also causing significant damage to the marine environment and the coastal areas, including the urban and economic growth of the affected coastal zones. The Exxon Valdez (1989), Atlantic

Empress (1979), Amoco Cadiz (1978), and Torrey Canyon (1967) (**Figure 1**) are among the most renowned oil spills in history, that alarmed the scientific community and directed all toward the management of anthropogenic environmental disasters in the marine environment.

Spill modeling has a long history [3]. Since 1960s, numerous oil spill models have been developed to simulate weathering processes and forecast the fate of oil

Figure 1.
(a) Exxon Valdez (1989); (b) Atlantic Empress (1979); (c) Amoco Cadiz (1978); (d) Torrey Canyon (1967).

spilled, in terms of providing valuable support to both contingency planners and pollution response teams. These models, developed by various organizations, companies, and researchers, vary in complexity, applicability to location, and ease of use [4]. There are two categories according to the Industry Technical Advisory Committee (ITAC) for oil spill response. One category, known as oil weathering models, estimates how oil properties change over time, but does not predict potential migration of the slick. The second category includes trajectory or deterministic models, stochastic or probability models, hind cast and three-dimensional models. In addition to predicting weathering profiles, these models estimate the evolution of a slick over time. The first-generation models were limited to tracing the movement of rigid bodies under the influence of current and wind advection. Oil spreading was justified by the balance of forces, and although chemical processes played significant role, most of them were ignored in this modeling [5]. Second-generation models have been summarized in [6, 7], where some of the principal forces and major phenomena, determining the behavior and the fate of the chemicals at sea, were analyzed. The recent generation models are reviewed in many papers [8–12]. Some of the oil spill models that are currently available are: General NOAA Operational Modeling Environment (GNOME), MEDSLIK-II, SeaTrackWeb (STW), Model Oceanique de Transport d' Hydrocarbures (MOTHY), DieCAST-SSBOM (Shirshov-Stony Brook Oil spill transport Model), COastal Zone OIL spill model (COZOIL), etc.

These operational computer models, pondering major physical processes, efficiently simulate the movement, the spreading and the evolution of the floating chemicals. Strong winds and significant water currents, which are contemplated, are major contributors to both the creation of accidents and faster spreading of the oil. Such models can be used either in hind cast mode capable of tracing the source of a spill, or in forecast mode predicting the path, the horizontal dispersion, and the mass balance, assisting in this way the real-time crisis management.

2. Oil weathering

From the time the oil spreads over the water, under the influence of hydrometeorological conditions (wave, winds, currents, solar radiation, etc.), oil properties (density, viscosity, pour point, etc.), and discharge characteristics (instantaneous, continuous, surface, depth), several processes take place, directly or indirectly related, which disperse the oil and change its properties [13, 14]. These time-dependent processes comprise physical, chemical, and biological ones, such as evaporation, dissolution, emulsification, dispersion, etc. (**Figure 2**). The combination of these processes is called oil weathering. In other words, weathering is a series of chemical and physical changes that cause spilled oil to break down and become heavier than water. Winds, waves, and currents may result in natural dispersion, breaking a slick into droplets which are then distributed throughout the water. These droplets may also result in the creation of a secondary slick or thin film on the surface of the water.

In general, weathering processes can be divided into three categories. Processes such as spreading, evaporation, dissolution, dispersion, and emulsification are rapidly occurring (within hours) and have immediate effects on the oil slick, as shown in **Figure 3**. Sedimentation, biodegradation and photo-oxidation operate more slowly (within months) and comprise the long-term mechanisms for the breakdown of hydrocarbons in the environment. Sedimentation, stranding, and oil-ice interaction are important processes, under distinct environmental conditions. Evaporation is probably one of the most significant processes that affect the surface

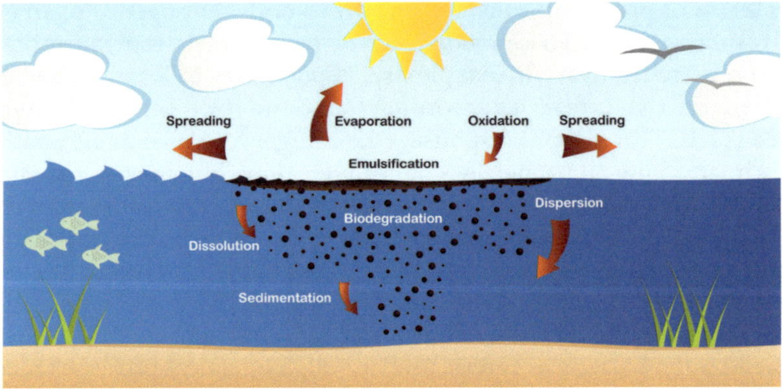

Figure 2.
Weathering processes that oil undergoes in the sea [15].

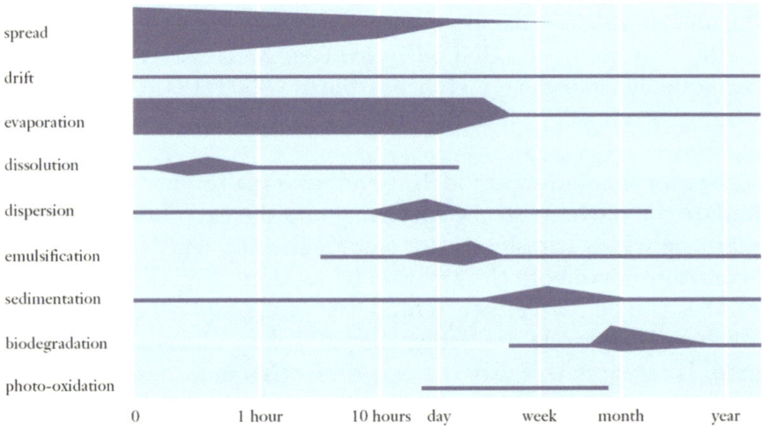

Figure 3.
Duration of oil slick processes in the sea.

oil particles, in sea or on coast, and therefore, it will play an important role in the construction of the models presented here.

In the following chart, the vertical axis portrays the physicochemical processes that oil is going through, and the time each one of them requires is depicted in the horizontal axis, whereas the line thickness denotes the most critical phase of each stage.

Natural processes may act to reduce the severity of an oil spill or accelerate the decomposition of spilled oil. More analytically, these natural processes are described below [15]:

- **Evaporation** occurs when the lighter substances within the oil mixture become vapors and leave the surface of the water. This process leaves behind the heavier components of the oil, which may undergo further weathering or may sink to the ocean floor. For example, spills of lighter refined petroleum-based products such as kerosene and gasoline contain a high proportion of flammable components known as light ends. These may evaporate completely within a few hours, thereby reducing the toxic effects to the environment. Heavier oils leave a thicker, more viscous residue, which may have serious physical and chemical impacts on the environment. Wind, waves, and currents increase both evaporation and natural dispersion.

- **Emulsification** is a process that forms emulsions consisting of a mixture of small droplets of oil and water. Emulsions are formed by wave action, and greatly hamper weathering and cleanup processes. Two types of emulsions exist: water-in-oil and oil-in-water. Water-in-oil emulsions are frequently called "chocolate mousse," and they are formed when strong currents or wave action causes water to become trapped inside viscous oil. Chocolate mousse emulsions may linger in the environment for months or even years. Oil and water emulsions cause oil to sink and disappear from the surface, which give the false impression that it is gone and the threat to the environment has ended.

- **Biodegradation** occurs when microorganisms such as bacteria feed on oil. A wide range of microorganisms is required for a significant reduction of the oil. To sustain biodegradation, nutrients such as nitrogen and phosphorus are sometimes added to the water to encourage the microorganisms to grow and reproduce. Biodegradation tends to work best in warm water environments.

- **Oxidation** occurs when oil contacts the water and oxygen combines with the oil to produce water-soluble compounds. This process affects oil slicks mostly around their edges. Thick slicks may only partially oxidize, forming tar balls. These dense, sticky, black spheres may linger in the environment, and can collect in the sediments of slow-moving streams or lakes or wash up on shorelines long after a spill.

3. Oil spill simulation models

A detailed presentation of the oil slick models developed by the Aristotle University of Thessaloniki, which successfully incorporates most of the aforementioned processes, is following. A concise description of the models' construction and their operational applications are briefly elaborated.

3.1 SOSM (Sea and Oil Slick) 3D Simulation Model

A model suite, comprising of a coastal hydrodynamic circulation, a wave generation model, and a Lagrangian (tracer) model for the transport and physicochemical evolution of an oil slick, is synthesized by the Aristotle University of Thessaloniki [16]. It follows a model that was developed in the context of an EU DG XI project, in cooperation with the UGMM Group of the Belgian Ministry of Public Health and Environment [17]. Following that, the Oil Spill Model (OSM), based on the PARCEL model, was coupled with the hydrodynamic Princeton Ocean Model (POM), and applied on the Navarino bay, in SW Greece, adjacent to the Ionian Sea, to monitor the pollution impact on the benthic structure and the related fisheries [18]. Parallel to that, the Poseidon pollutants transport model (PPTM) was developed to provide real-time data and forecasts for marine environmental conditions and protect the Greek seas from oil spills, as an operational management tool, which consists of a 3D floating pollutant prediction model coupled with a weather, a hydrodynamic, and a wave model [19]. PPTM was applied in four areas of strategic interest in the Greek seas.

The two basic components of the model, that is, the hydrodynamic and the oil transport, are briefly presented here, followed by the applications under the prevailing circulation conditions in a coastal basin of Greece in the NW Aegean Sea, where the second largest port of the country is located in the town of Thessaloniki, with dense maritime traffic [16].

3.1.1 Model construction

The model is composed of two parts: the hydrodynamic part, comprising of the circulation and the wind wave subparts, and the oil slick transport part. The first is aimed at the description of the 3D nearly horizontal flow velocity vector field due to wind and/or tide. It is written with respect to the free surface ζ(x,y,t) and the depth-varying velocity components u(x,y,z,t), v(x,y,z,t)

$$\frac{\partial \zeta}{\partial t} + \frac{\partial}{\partial x}\int_h u dz + \frac{\partial}{\partial y}\int_h v dz = 0 \tag{1}$$

$$\frac{Du}{dt} = -g\frac{\partial \zeta}{\partial x} + N_h\left(\frac{\partial^2 u}{\partial x^2} + \frac{\partial^2 u}{\partial y^2}\right) + N_v\frac{\partial^2 u}{\partial z^2} = 0 \tag{2}$$

$$\frac{Dv}{dt} = -g\frac{\partial \zeta}{\partial y} + N_h\left(\frac{\partial^2 v}{\partial x^2} + \frac{\partial^2 v}{\partial y^2}\right) + N_v\frac{\partial^2 v}{\partial z^2} = 0 \tag{3}$$

The horizontal eddy viscosity-diffusivity N_h is deduced from the 4/3 law with a characteristic length related to the used Dx, and the vertical eddy viscosity-diffusivity N_v is controlled by the surface friction velocity, the mixing length, the wave height and period. The model is solved numerically by finite differences, on a staggered horizontal Arakawa C grid (**Figure 4**), and a vertical discretization, following the σ-coordinates transform, using the technique of fractional steps (i.e., resolving explicitly for the horizontal advection and implicitly for the vertical momentum diffusion) [20].

The wind waves are resolved by an empirical model WACCAS [21, 22] based on the JONSWAP spectrum. The wave height, wave period, and wave direction on the center of the mass of the slick is used for the computation of the Stokes' drift and the vertical viscosity-diffusivity. The oil slick simulation and tracking are done in Lagrangian coordinates [23, 24], by replacing the oil mass by a big number (10^4) of particles or parcels, each one of which represents a group of oil droplets of similar size and composition. The oil spill simulation by a large number of passive (not

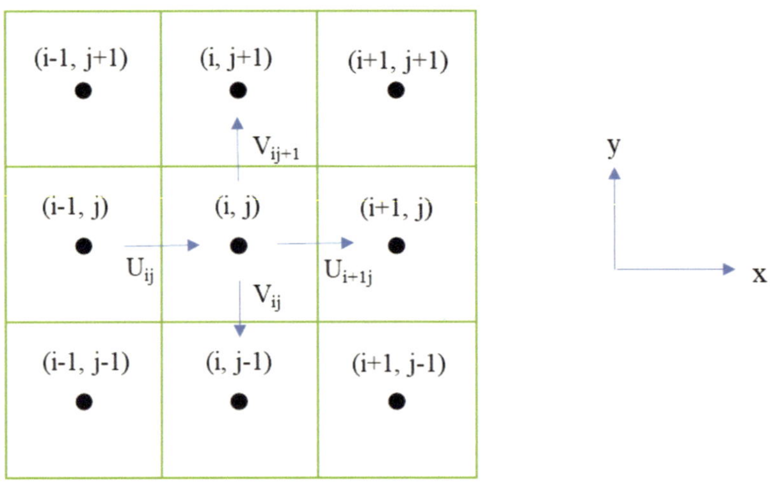

Figure 4.
Arakawa C staggered grid configuration.

interacting with the existing hydrodynamics) particles, transported in a Lagrangian frame of reference, is the most effective procedure, used in most contemporary models as: (a) it can describe oil spill geometries and shapes at subgrid level, (b) the model results are not distorted by numerical errors, like numerical diffusion, and (c) the fraction of the initially introduced particles (at the spill source) reaching the coast can be considered as the probability of pollution from a minor spill [25]. According to the tracer technique, it is theoretically possible to track the motion of all the "particles" (parcels) of an oil plume as they are transported by the sea. The density, the buoyancy velocity, the volatility, and other properties are statistically distributed to the particles. Their trajectories are tracked in space and time. Each one is advected according to the local (interpolated) water velocity and it is diffused by means of random walks related to the local diffusivity induced by currents and waves.

Evaporation and emulsification are processes quantified by means of commonly used functions relating the processes to the oil properties (to the fraction density, the temperature, the vapor pressure, the wind and water velocity, etc.). The Mackay approach is used [26, 27] for the computation of the evaporated volume fraction from each oil component. Emulsification is activated for oil fraction with densities and under wave height/length values beyond critical ones. The evaporation and emulsification processes result in: (a) loss of contained oil volume in the evaporated particles and (b) creation of a mixture of water in oil and formation of a floating mousse. Another important oil weathering process accounted in this model is beaching, which is better described by the duration of trapping on the coastal boundary of a beached particle as a function of the coastal morphology (from rocky to flat sandy beach). This process gives crucial information on the oil quantities retained on the coast and the consequent environmental effects.

The spatial and temporal distributions of values like oil concentration (or volume per unit area) are deduced from the number of particles and the contained volume inside each mesh of the discretization grid used for the hydrodynamic analysis. It is worth noting that the time step Dt, used for the integration of the hydrodynamic model, differs by orders of magnitude from the one used for the tracking of the oil slick.

3.1.2 Operational application

The produced operational code was updated and adapted to the environment of a coastal basin of NW Aegean Sea, namely Thermaikos bay. The commercial port of Thessaloniki hosts among others a busy oil terminal. Although the terminal operates under the strict International Maritime Organization (IMO) and European Union (EU) regulations for oil terminal operational safety, the port area is at risk for a potential oil slick accident.

The transport and fate of the slick are closely related to the hydrodynamic conditions in the bay, the bay morphology, and the oil properties. The hydrodynamic circulation in the bay is experiencing mainly N-NW winds, whereas the circulation patterns are regulated by the bay bathymetry and the coastal topography. The bay configuration and bathymetry are shown in **Figure 5**, along with the initial oil spill location.

The model was tested for hypothetical scenarios of oil spill under NW and NE winds of 45 knots, to assess the risk of pollution in the coastline of the touristic areas of Chalkidiki (east) and Katerini (west), respectively. The oil is presumably spilled in the port of Thessaloniki (**Figure 5**), as a result of a hypothetical accidental spill during a tanker operation. An oil spill of 1000 parcels, representing a total volume of 10,000 tn, is released instantaneously on the sea surface. Total simulation run

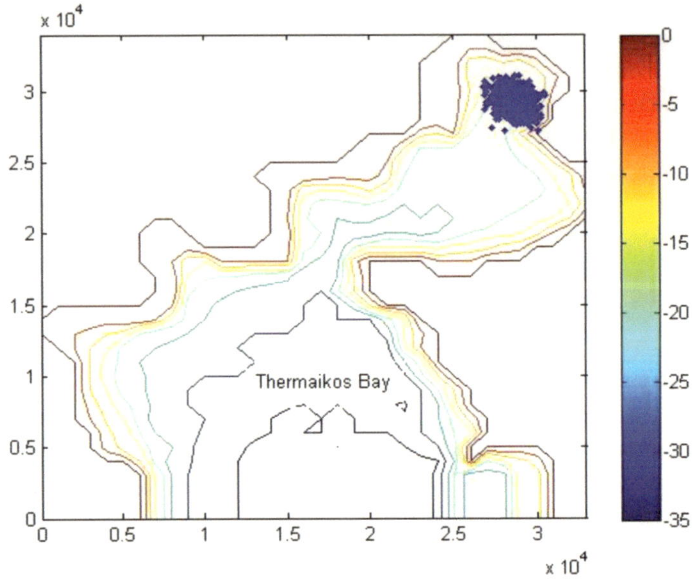

Figure 5.
Thermaikos bay configuration and bathymetry, and initial oil slick conditions.

was for 72 hrs (3 days), providing output results every 3 hrs. The evolution of the slick trajectories, for the first 24 hrs and for the two wind-generated circulation scenarios, is depicted in **Figures 6** and 7, respectively, documenting the difference in the coastal impact of the slick for the two cases.

In the case of strong NW wind, the touristic coast opposite to the bay is damaged within some hours after the spillage, while in the case of the NE wind, the oil slick travels along the bay toward the open sea giving time to the relevant authorities for intervention, blocking, and final cleaning. The dispersion effect is also revealed showing that the oil slick diameter is moderately increasing during this time.

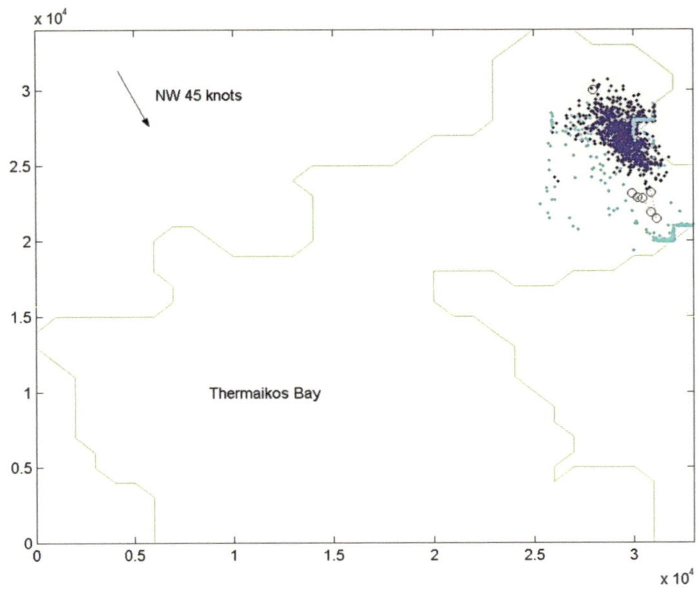

Figure 6.
Evolution of the oil slick under strong NW wind.

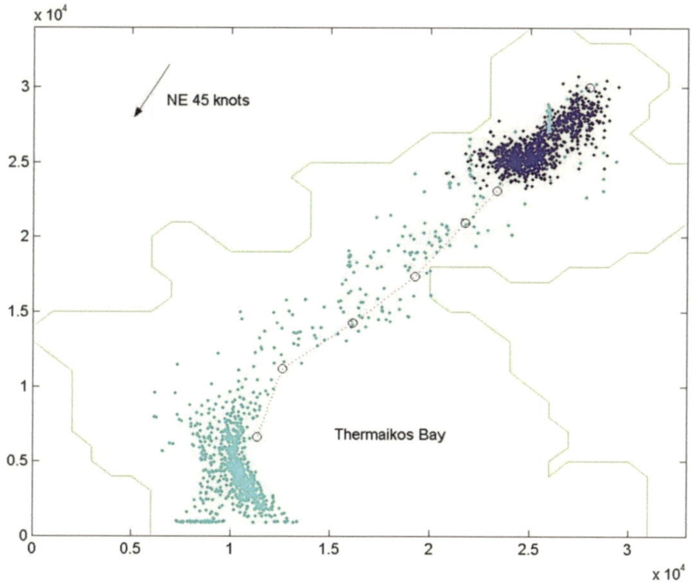

Figure 7.
Evolution of the oil slick under strong NE wind.

Figure 7 depicts more clearly the spreading of the oil spill and its statistical center of gravity movement (noted by red circles), as it is advected for many kilometers in Thermaikos bay. The oil slick dispersion after 3 hrs is shown in dark blue color, whereas the spreading of the plume at the end of a day (after 24 hrs) is in light blue. The spiral movement of the oil's statistical center of gravity in **Figure 6**, near the shore, may be attributed to the inertial currents frequently observed in the sea, and points out that the largest part of the oil reaches the shore (beaching) within the initial time steps (after 9 hrs). The model results show that a contingency emergency plan should be, in this case, activated in less than 12 hrs, in order to avoid the social, economic, and environmental damage in the area.

3.2 DIAVLOS-NASOS

The universities of Thessaloniki and Athens cooperated, in the framework of the project DIAVLOS [28–30], to produce a simulation model that could trace an oil spill in case of an accident in the Alexandroupolis Gulf, which is considered of high risk, after the construction of the Burgas-Alexandroupolis oil pipeline that has been signed by the Greek, Bulgarian, and Russian governments (on March 15, 2007), and is expected to carry huge oil quantities (35–50 Mtn/yr) that will be loaded to tankers and further transported by sea to international markets.

3.2.1 Model construction

A 48-hr oil spill dispersion forecasting system was developed and implemented in the wider northern Aegean Sea. The system was based on wind, wave, and ocean circulation models, coupled with the operational systems ALERMO and SKIRON of the Department of Physics of the University of Athens, and an oil spill dispersion model by the Department of Civil Engineering of the Aristotle University of Thessaloniki. The basic components of the model are presented hereafter.

3.2.1.1 Meteorological model

The meteorological model is part of the weather forecasting system SKIRON developed by the University of Athens, which is based on a Limited Area Model of Eta/NCEP, which provides high-resolution (of 1°) weather forecasts for 120 hrs [31]. In this work, the non-hydrostatic model Eta of the operational forecasting system SKIRON has been used, with an even higher resolution (of 1°/10) since the description of non-hydrostatic phenomena is necessary for the model to be applied for such a high resolution. The available data are: air temperature and humidity at 2 m, wind velocity and direction at 10 m, short- and long-wave radiation, atmospheric pressure at the sea level, and precipitation.

3.2.1.2 Hydrodynamic model

The hydrodynamic forecasting model, which was developed for the area under study, is based on the well-known numerical model Princeton Ocean Model (POM) [21], a model that has been widely used for the simulation of open and coastal sea circulation. The Department of Physics at the University of Athens developed a POM-based sea circulation forecasting model, named ALERMO, under the European programs MFSPP and MFSTEP, which was made public in 2004 (http://www.phys.uoa.gr). Its resolution is 2 nm and it is coupled with the Mediterranean forecasting system MFS-OGCM (INGV, Italy), which also assimilates satellite data and field measurements (XBTs, CTDs, Profiling Float). ALERMO uses an innovative variational initialization technique (VIFOP) for its initial conditions and provides initial and boundary conditions to other coastal forecasting systems (e.g., Greece, Cyprus, and Israel). The developed hydrodynamic model has been applied to the area of interest (North Aegean: 38.7°–41.1° N, 22.5°–27.1° E) with horizontal resolution of 1°/60 (1 nm) and 25 σ-layers in the vertical. The open boundary conditions, as well as the initial conditions, are provided by ALERMO. Finally, the momentum, heat and water fluxes at the air-sea interface are calculated by using the weather forecasting data provided by the meteorological model. A series of sensitivity tests have been conducted, with respect to the atmospheric conditions and the inflow of water from the Black Sea [32].

3.2.1.3 Wave model

The wave model WAM [33, 34] has been adapted and applied to the area under study, with a resolution of 1°/20 by the University of Athens. This model is a third-generation numerical model, which has been widely used, as it best describes the evolution of the wave spectrum in the sea. WAM makes a distinction between deep and shallow waters, depending on the sea depth at the area, where the wave equations are being applied.

3.2.1.4 Oil spill dispersion model

The oil spill transport and dispersion model, named NASOS (North Aegean Sea and Oil Slick), is based on the 3D model that has been developed by the civil engineering department of the Aristotle University of Thessaloniki [16, 17]. The produced operational code was updated and adapted to the area of interest (38.7°–41.1° N, 22.5°–27.1° E), that is, the North Aegean Sea. This model assumes that the oil slick is described by a big number (equal to 10^4) of particles (or "parcels"), each of which represents many cubic meters of oil, and has individual physicochemical properties. The distinction of the oil spill in these "parcels" is supported by the fact that oil contains a variety of hydrocarbon components with different physicochemical properties, and therefore makes the monitoring of the evolution of each "parcel" easier. This model recognizes four different oil types, enough to cover the variety of physicochemical properties and their effect on the environment. In terms of the simulation of the fate and transport of the oil spill in the sea, the model recognizes the following processes: initial spreading and transport, horizontal and vertical diffusion and dispersion, evaporation, emulsification, beaching, and sedimentation. The Mackay approach [26, 27] is again used for the computation of the evaporated volume fraction from each oil component. Emulsification is activated for oil fraction with densities and wave height/length ratios beyond critical values. The evaporation and emulsification processes lead on the one hand to the deficit of contained oil in the evaporated particles, and on the other hand to the creation of a mixture of water and oil, and formation of a floating mousse. Beaching, which expresses the entrapment of oil on the beach, is expressed by the duration of particles trapped on the coastal boundary. It strongly depends on the coastal morphology (rocky to flat sandy beach).

3.2.2 Operational application

This study's objective was to produce an operational tool to forecast the fate of a possible oil spill in Alexandroupolis gulf and contribute to manage environmental crises from such pollution, after the Burgas-Alexandroupolis pipeline would be constructed. It was clearly very important to make this oil spill dispersion forecasting system user-friendly and easily accessible by all the involved authorities (e.g., the Ministry of Mercantile Marine/Oil Spill Prevention Department, Prefecture of Evros, local port authorities, etc.).

The required input data are the initial time and location of the oil spill, the oil quantity, and quality characteristics (oil type). These data can be introduced to

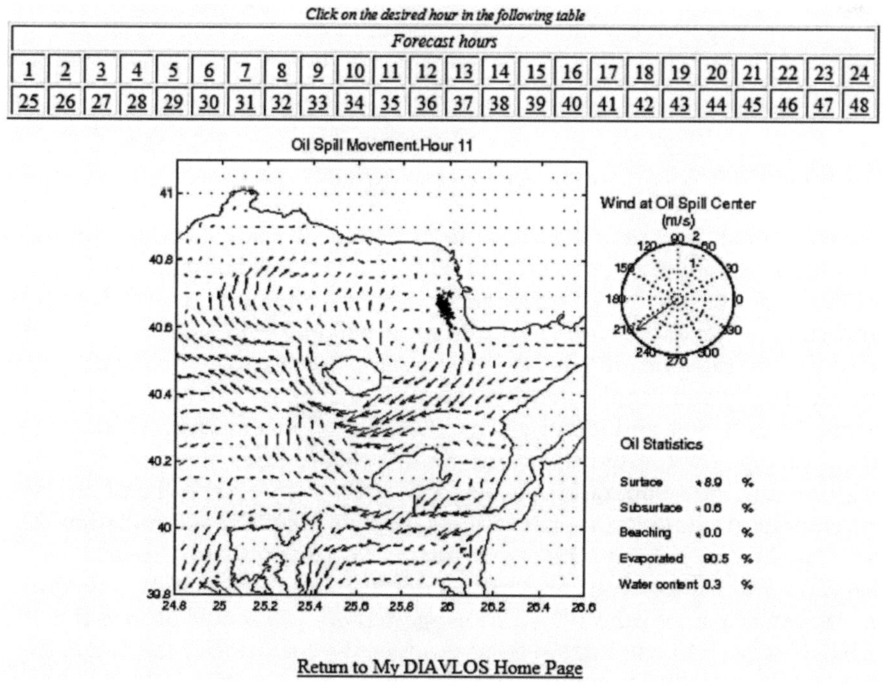

Figure 8.
Graphical and statistical results of the oil spill operational transport model (DIAVLOS).

DIAVLOS forecasting system's Website on-line by the user. The information is processed by the wave-atmospheric-hydrodynamic forecasting system. The oil spill dispersion model runs for 48 hrs and provides hourly images in less than 3–4 minutes. The system can also provide an updated forecast if updated input information is available (in less than 48 hrs). The model computes the hydrodynamic flow and the 3D oil spill evolution and returns to the user the following information (in percentages): surface and subsurface dispersion, evaporation, emulsification (or water content), and beaching of the oil. The output of the model, as **Figure 8** shows, contains a graphical representation of the hypothetical oil spill accident area, as well as the horizontal dispersion of the oil, depicted by different colors (surface, subsurface, or beached). Additionally, the University of Athens provides 60-hr high-resolution atmospheric and oceanic forecasting at the region, to facilitate operations to contain oil spill spreading, and beaching and cleaning operations. Wind, wave, and oceanic circulation forecasts are available on DIAVLOS Website by the Ocean Physics and Modelling Group and the Atmospheric Modelling and Weather Prediction Group of the University of Athens. The password-protected and interactive Website (http://diavlos.oc.phys.uoa.gr) was created and offered to public in May 2008, combining efficiency and simplicity.

The system was also tested against field observations of special drifting floats that monitor the fate and evolution of oil spills. These "smart drifters," were especially designed for this project by MARAC Electronics Co. (Greece) in cooperation with the University of the Aegean, co-flowing and transmitting along with the oil slick. The model proved to give satisfactory results and in most cases the forecasting error is quite small, allowing the operational use of the system.

3.3 BSB Net-Eco—SLICKNEW

This latest study utilizes another upgraded version of PARCEL and 3D Sea and Oil Slick (SOSM) models, developed by the Aristotle University of Thessaloniki [16, 17]. The newly formalized model was tested in the Aegean Sea [35] and adjusted to the characteristics to the Black Sea, and more particularly to the Azov Sea [36].

3.3.1 Model construction

The model suite comprises of a Lagrangian (tracer) model for the transport and physicochemical evolution of an oil slick [37]. The input requirements of the model are surface wind velocities, air temperature, vertical and horizontal diffusion coefficients, surface currents, wave characteristics (height, period, direction), and, in terms of the oil transport, the initial coordinates of each parcel, its initial volume and mean density and droplet diameter, as well as evaporation rate and parameters relevant to the oil type and identity. The POSEIDON system [19], which is utilized in this modeling effort, provides information on wind speed and direction, atmospheric pressure, air temperature, wave parameters, current speed and direction, water temperature, salinity, dissolved oxygen, chlorophyll-a, and radiation. The observational basis of the system is a network of oceanographic 11 buoys (7 SEAWATCH, 3 Wavescan and 1 deepwater SEAWATCH module), operating in the Aegean Sea since June 1999. The suggested oil spill model utilizes the POSEIDON wave and wind datasets, to produce the sea velocity fields due to currents and waves. The wave, generated near surface, velocity field is computed from the classical Stokes' drift formulae [20], based on the local values of wave height H_s and wave period T_s.

$$U_m(z) = \left(\frac{\pi \cdot H_s}{L_0}\right)^2 \frac{C_0}{2} \frac{\cosh\left(2k(H_s + z)\right)}{\sinh\left(k \cdot H_s\right)^2} \tag{4}$$

where $L_0 = g \cdot T_s^2/2\pi$, $C_0 = g \cdot T_s/2\pi = L_0/T_s$, and $k = 2\pi/L_0$ (π = 3.14). The horizontal diffusion coefficient D_h is estimated adopting the Smagorinsky formula [21, 38] as follows:

$$D_h = C \cdot dx \cdot dy \cdot \left[\left(\frac{\partial u}{\partial x}\right)^2 + 0.5\left(\frac{\partial v}{\partial x} + \frac{\partial v}{\partial y}\right)^2 + \left(\frac{\partial v}{\partial y}\right)^2\right]^{0.5} \tag{5}$$

where C the horizontal diffusion coefficient. This equation is used to estimate the velocity of each particle, if it is selected via a random number from a sample following the uniform distribution over a range {−Ur, +Ur}

$$U_r = \sqrt{\frac{6 \cdot D_h}{\Delta t}} \tag{6}$$

The generation of that sample of random velocity values is based on Monte Carlo sampling, a very common and powerful procedure in simulation theory [37]. The vertical displacement of the oil "particles" is considering the vertical diffusion due to currents and waves [39]:

$$D_V = D_{Vc} + D_{Vw} \tag{7}$$

$$D_{Vc} = \lambda \cdot h \cdot W \tag{8}$$

$$D_{Vw} = 0.028 \cdot \frac{H_s^2}{T_s} \cdot e^{\frac{4\pi z}{L_0}} \tag{9}$$

where λ = 0.001; h the water depth; and $W = \sqrt{W_x^2 + W_y^2}$ for W_x, W_y the given wind velocities in the x, y axes, respectively. Consequently, the vertical velocity due to buoyancy and diffusion of the oil "particles" is given, similar to the horizontal velocity [37], by

$$W_r = \sqrt{\frac{6 \cdot D_V}{\Delta t}} \tag{10}$$

With respect to the input requirements of the slick model, it also requires bathymetry data of the selected area, and a file containing the characterization of the coastal meshes, according to their oil-holding capacity and the open sea boundaries. Based on all that, new horizontal positions of the oil "particles" are estimated; some are "trapped" on the beach, others may be vertically displaced due to buoyancy and diffusion; a fraction of heavy classes of oil may be emulsified over a certain wave curvature, whereas a light oil fraction in sea or on coast may be evaporated. Among the oil weathering processes that take place in the sea is evaporation. It affects the surface oil particles, in sea or on coast. A complete review of various approaches in estimating the evaporated oil is presented in [40]. Thus, adopting the empirical equation of oil evaporation representative of the oil type "Barrow Island, Australia," the evaporation formula used in this model is described by the following equation:

$$\%Ev = (4.67 + 0.045 \cdot T)\ln(t) \tag{11}$$

where T is the air temperature (°C) and t the time (in minutes). Another important oil weathering process is emulsification, which is expressed as a function of the wind velocity W expressed as $W = \sqrt{W_x^2 + W_y^2}$ and the temperature T [26]:

$$Em = \frac{1 - e^{-0.0000056\left(1+W^2\right)\cdot T}}{1.25} \tag{12}$$

Different oil products react differently to these processes. Lighter oil fractions tend to evaporate, whereas heavier fractions tend to emulsify. Therefore, the model, taking that into account, can simulate different oil types according to their density and buoyancy velocity. The processes of photo-oxidation and biodegradation are not considered in this version of the model, as their effects are more significant at a later stage of a spill's evolution (see **Figure 3**). All particles are considered to spread at a single location, while they can be released all at the same time (instantaneous discharge), or in sequence over a given period of time (continuous discharge of specified duration).

3.3.2 Operational application

The produced operational code was adapted to the environment of a coastal basin of NW Aegean Sea (39.96°–40.66° N, 22.50°–23.40° E), namely Thermaikos bay, which also includes Thessaloniki's gulf [41]. The area that has been selected to be studied is shown in **Figure 9** [42], with a horizontal discretization of 1°/30

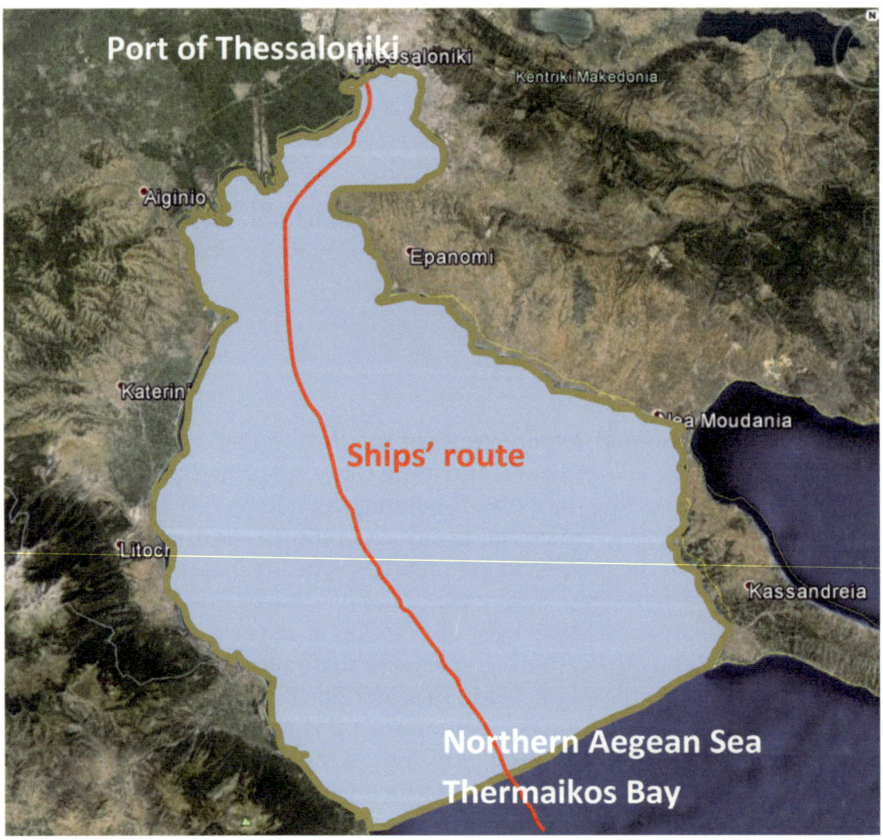

Figure 9.
Thermaikos bay and ships' route to and from the port of Thessaloniki.

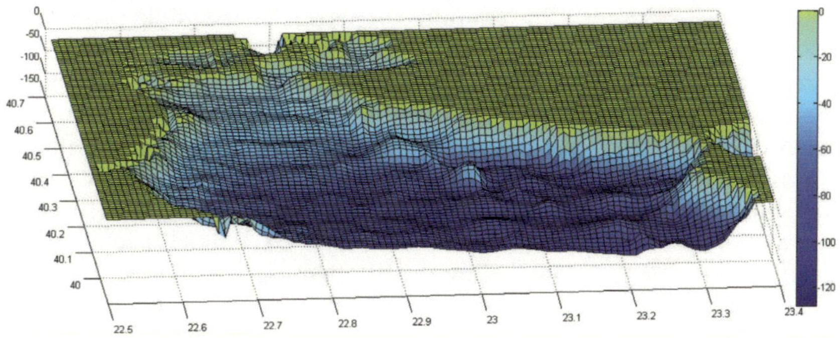

Figure 10.
Thermaikos bay bathymetry.

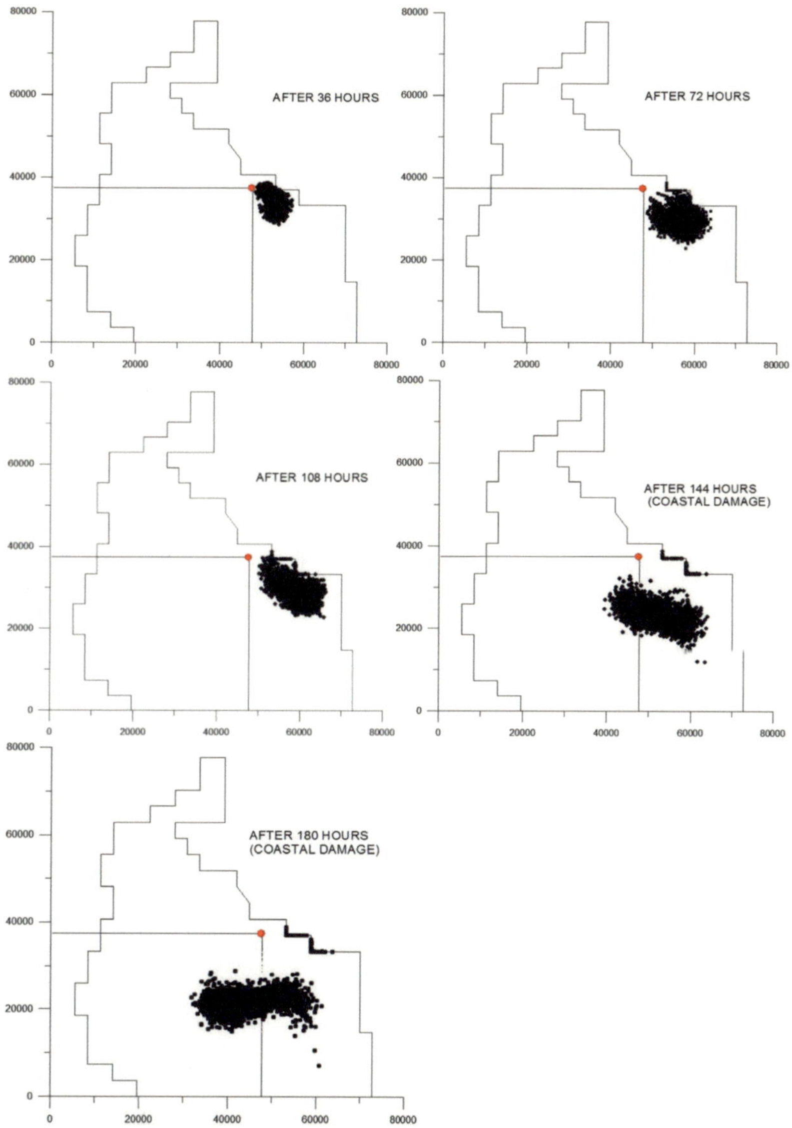

Figure 11.
Snapshots of the spatiotemporal evolution of an oil spill hypothetically released near the coast of Chalkidiki, at the east side of Thermaikos bay, after: (a) 36 h, (b) 72 h, (c) 108 h, (d) 144 h, and (e) 180 h.

(Dx = 2800 m, Dy = 3700 m). The commercial port of Thessaloniki provides a good study area, as it is enclosed in Thermaikos bay with an important coastline and a busy oil terminal. Although the terminal operates under strict IMO (International Maritime Organization) and EU (European Union) regulations for oil terminal operational safety, the port area is a potential oil slick accidental source [38].

The transport and fate of the slick are closely related to the hydrodynamic conditions in the bay, the bay morphology and the oil properties. The hydrodynamic circulation in the bay is sustained mainly by the strong NW to NE winds and the circulation patterns are regulated by the bay bathymetry and the coastal topography (**Figure 10**). Following a thorough consideration, the period of data selected to be used from the POSEIDON dataset was a winter week of January (21–28/01/2014) which exhibited significant winds, able to transport the oil slick and produce substantial results. The data were given for 29 time steps (6 hrs. each). The POSEIDON dataset for the Aegean Sea offers 3D hydrodynamic forecasts (POM) nested to POSEIDON Med forecast, for the depth levels of 5, 15, 30, 50, 80, and 120 m. The depth in the area under study, as shown in **Figure 10**, is less than 100 m. One hypothetical oil spill scenario has been selected to portray the model's simulation capabilities and results, at the east side of Thermaikos bay, near the touristic areas of Chalkidiki (**Figure 11**). An oil spill of 5000 parcels, representing an estimated total volume of 10,000 tn, was hypothetically released instantaneously on the sea surface. Total simulation run was for 180 hrs (7.5 days). **Figure 11** presents snapshots of the spatiotemporal evolution of the oil slick at a 36-hr step after the release. The touristic coast of Chalkidiki is damaged within some hours after the spillage (clearly after 2–3 days). The dispersion effect is also revealed showing that the oil slick diameter is moderately increasing during this time.

Acknowledgements

Part of this research has been funded by the General Secretariat of Research and Technology (GSRT) of Greece, under the title "Development and application of crisis management tools aiming at the prevention from oil spill pollution in the area of Burgas – Alexandroupolis pipeline (N.E. Aegean Sea) – DIAVLOS." Another part of this research has been produced with the assistance of the European Union, through the BSB JOP 2013-2017 (under the project with the acronym "BSB Net-Eco").

Thanks

I would like to express my gratitude to all my colleagues who contributed to this work, my co-authors to the published articles and conference papers, but most of all to Emeritus Professor Ch. Koutitas.

Author details

Antigoni Zafirakou
Aristotle University of Thessaloniki, Thessaloniki, Greece

*Address all correspondence to: azafir@civil.auth.gr

IntechOpen

© 2018 The Author(s). Licensee IntechOpen. This chapter is distributed under the terms of the Creative Commons Attribution License (http://creativecommons.org/licenses/by/3.0), which permits unrestricted use, distribution, and reproduction in any medium, provided the original work is properly cited.

References

[1] NOAA. Available from: https://oceanservice.noaa.gov/facts/pollution.html [Accessed: 17-06-2018]

[2] Themeli S, Tsami E. Multi-parametric analysis of the mitigation plans for the oil spill pollution in the sea [thesis]. Greece: Dept. Civil Engineering, Aristotle University of Thessaloniki; 2017

[3] Kulygin VV. A review of modern approaches of oil spill modeling. In: Proc. of International Conference on Oil and gas of arctic shelf; 2006

[4] Kubryakov A, Korataev G. A short comparison of oil spill drift forecasting systems, with an overview of the SEATRACKWEB system for the Moninfo Project. Commission on the Protection of the Black Sea Against Pollution. Instabul, Turkey: Permanent Secretariat; 2010

[5] Fay JA. The spread of oil slicks on a calm sea. In: Hoult DP, editor. Oil on the Sea. New York: Plenum Press; 1969

[6] Huang JC, Monastery FC. Review of the state of the art of oil spill simulation models; Final Report. Washington DC: American Petroleum Institute; 1982

[7] Huang JC. A review of the state of the art of oil spill fate/behavior models. In: Proceedings International Oil Spill Conference; February 28–March 3; San Antonio, TX, USA. Washington, DC: American Petroleum Institute; 1983. pp. 313-322

[8] Volckaert F, Tombroff D. Progress Report to EEC Div. XI/B/l.; 1989

[9] ASCE (American Society of Civil Engineers). State-of-the-art review of modeling transport and fate of oil spills. Task Committee on modeling of oil spills. ASCE. Journal of Hydraulic Engineering. 1996;**122**:594-609

[10] Reed M, Johansen O, Brandvik P, Daling P, Lewis A, Fiocco R, et al. Oil spill modelling towards the close of 20th century: Overview of the state of the art. Spill Science & Technology Bulletin. 1999;**5**(1):3-16

[11] Etkin DS, Michel J, McCay DF, Boufadel M, Li H. Integrating state-of-the-art shoreline interaction knowledge into spill modeling. In: International Oil Spill Conference; 2008. pp. 915-922

[12] De Dominicis M, Pinardi N, Zodiatis G, Lardner R. MEDSLIK-II, a Lagrangian marine oil spill model for short-term forecasting—Part I: Theory. Geoscientific Model Development. 2013;**6**:1851-1869

[13] Wheeler RB. The fate of petroleum in the marine environment; Special Report. Houston, USA: Exxon Production Research Co.; 1978. p. 124

[14] API (American Petroleum Institute). Fate of Spilled Oil in Marine Waters; Publication Number 4691; 1999

[15] ITOPF (International Tank Owners Pollution Federation) TIP 02: Fate of Marine Oil Spills. Available at: http://www.itopf.org/knowledge-resources/documents-guides/document/tip-02-fate-of-marine-oil-spills/ [Accessed 20-07-2018]

[16] Zafirakou-Koulouris A, Dermisi SC, Koutitas C, Dermissis V. Modeling the evolution and fate of an oil slick with a 3-D simulation model. In: Proc. of Protection and Restoration of the Environment IX. Cephalonia, Greece; 2008

[17] Koutitas C. Joint Research Project PARCEL. Final Report. EU DG for the Environment Div. XI/B/1. Thessaloniki; 1991

[18] Petihakis GI, Triantafyllou GN, Koutitas CG. Prediction and prevention of oil contamination and monitoring of the benthic structure and related fisheries in connection with the pollution impact. Systems Analysis Modelling Simulation. 2001;**41**:169-197

[19] Pollani A, Triantafyllou G, Petihakis G, Nittis K, Dounas C, Koutitas C. The POSEIDON operational tool for the Prediction of Floating Pollutant Transport. Marine Pollution Bulletin. 2001;**43**(7–12):270-278

[20] Koutitas C. 3-D Coastal Circulation Models, an Engineering Viewpoint. In: Heaps N, editor. Three-Dimensional Coastal Ocean Models. American Geophysical Union; 1985

[21] Blumberg AF, Mellor GL. A Description of a three-dimensional coastal ocean circulation model. In: Coastal and Estuarine Science. Vol. 4. Washington D.C: AGU; 1987. pp. 1-16

[22] Burrows R. Recent wave hindcasting studies using the WACCAS model. In: Schrefler, Ziekniewicz, editors. Comp. Modelling in Ocean Engineering. Blakena; 1988. pp. 181-188

[23] Johansen V. Particle in Fluid Model for Simulation of Oil Drift and Spread. Trondheim, Norway: Oceanographic Center, SINTEF Group. 1985. p. 40

[24] Elliott A. Shear diffusion and the spread of oil in the surface layers of the North Sea. Deutsche Hydrografische Zeitschrift. 1986;**2**(39):113-137

[25] Samuels W, Amstutz D, Bahadur R, Ziemniak C. Development of a global oil spill modeling system. Earth Science Research. 2013;**2**:1-10

[26] Mackay D, Buist I, Mascarenhas R, Paterson S. Oil spill processes and models. Technical Report EE-8. Environment Canada Ottawa, Ontario; 1980. p. 17

[27] Stiver W, Mackay D. Evaporation rate of spills of hydrocarbons and petroleum mixtures. Environmental Science & Technology. 1984;**18**(11): 834-840

[28] Sofianos S, Kallos G, Mantziafou A, Tzali M, Zafeirakou A, Dermisis V, Koutitas Ch, Zervakis V. Oil-spill dispersion forecasting system for the region of installation of the burgas-alexandroupolis pipeline outlet (N.E. Aegean) in the framework of "DIAVLOS" Project. In: 9th Symposium on Oceanography & Fisheries, Proceedings; Vol. I. 2009

[29] Zafirakou-Koulouris A, Koutitas C, Sofianos S, Mantziafou A, Tzali M, Dermissi S. Oil spill forecasting with the aid of a 3D simulation model. Journal of Physical Science and Application. 2012;**2** (10):448-455

[30] Sofianos S, Zafirakou-Koulouris A, Koutitas C, Mantziafou A, Tzali M, Dermisi S. Oil spill dispersion forecasting with the aid of a 3D simulation model. Journal of Physical Science and Application. 2012;**10**: 448-453

[31] Kallos G, Koroni V, Lagouvardos K. The regional weather forecasting system SKIRON: An overview. In: Proceedings of the Symposium on Regional Weather Prediction on Parallel Computer Environments. Greece: University of Athens; 1997. pp. 109-122

[32] Tzali M. Study of the effect of water inflow of the Black Sea in the North Aegean Sea, with the use of arithmetic models, MS Thesis in Oceanography. Physics Department, University of Athens; 2007

[33] Komen GJ, Cavaleri L, Donelan M, Hasselmann K, Hasselmann S, Janssen PAEM. Dynamics and Modelling of Ocean Waves. Cambridge University Press; 1994

[34] WAMDI Group. The WAM model —A third generation ocean wave prediction model. Journal of Physical Oceanography. 1988;**18**:1775-1810

[35] Zafirakou A, Palantzas G, Samaras A, Koutitas C. Oil spill modeling aiming at the protection of ports and coastal areas. Environmental Processes. 2015;**2** (Suppl. 1):S41-S53

[36] Palantzas G, Zafeirakou A, Samaras A, Karambas T, Koutitas C. The use of oil spill trajectory models in oil pollution incidents' response. In: 11th Panhellenic Symposium Oceanography and Fishing, Mytilini, Lesvos, Greece; 13-17/05/2015; 2015

[37] Koutitas CG. Mathematical Models in Coastal Engineering. London, GB: Pentech Press; 1988

[38] Palantzas G, Koutitas C, Samaras A. The use of numerical simulation of oil spills in developing a chemical pollution contingency plan in ports. The case of the port of Thessaloniki, Greece. In: 13th International Symposium on Environmental Pollution and its Impact on Life in the Mediterranean Region, Thessaloniki; 8-12 October; MESAEP; 2005

[39] Koutitas C. Introduction to Coastal Engineering and Harbor Works. Ziti ed. Thessaloniki, Greece; 1998

[40] Fingas MF. Modeling oil and petroleum evaporation. Journal of Petroleum and Science Research. 2013;**2** (3):104-115

[41] Zafirakou A, Palantzas G, Samaras A, Koutitas C. The use of oil spill simulation in developing and applying oil pollution contingency plans in ports and coastal areas. In: Proc. of Protection and Restoration of the Environment XII; Skiathos, Greece; 2014

[42] Google Earth Map of Thermaikos Bay with Editing from the Authors. © 2014 Google, Data SIO, NOAA, U.S. Navy, NGA, GEBCO, Image Landsat, Imagery date: 4/10/2013, [Accessed on 17/07/2014]

Chapter 8

Response of Benthic Foraminifera to Environmental Variability: Importance of Benthic Foraminifera in Monitoring Studies

Maria Virginia Alves Martins, Cintia Yamashita, Silvia Helena de Mello e Sousa, Eduardo Apostolos Machado Koutsoukos, Sibelle Trevisan Disaró, Jean-Pierre Debenay and Wânia Duleba

Abstract

Foraminifera are eukaryotic unicellular microorganisms inhabiting all marine environments. The study of these protists has huge potential implications and benefits. They are good indicators of global change and are also promising indicators of the environmental health of marine ecosystems. Nevertheless, much remains to be learned about foraminiferal ecology. The goals of this chapter are (1) to provide a few examples from foraminifera studies, presenting possible use of foraminifera as bioindicators for the monitoring of transitional and marine ecosystems and (2) to highlight the importance of applying these organisms in environmental monitoring studies. A semienclosed coastal lagoon (Aveiro Lagoon; Portugal), an estuarine system (São Sebastião Channel; SE Brazil), a continental shelf sector (Campos Basin; SE Brazil), and a segment of continental slope (Campos Basin; SE Brazil) are used as examples.

Keywords: foraminifera, marine pollution, marine ecosystems, environmental stress, bioindicators

1. Introduction

The anthropic activity has been causing more and more negative effects on nature, among which includes the discharge of several types of pollutants from either domestic or industrial sources. The final destination for a majority of the pollutants are frequently the coastal areas where the pollutants may cause deleterious effects as harm to living resources and marine life, hazards to human health, hindrance to marine activities, including fishing and other legitimate uses of the sea [1]. Persistent pollutants like polycyclic aromatic hydrocarbons (PAHs),

IntechOpen

polychlorinated biphenyls (PCBs), pesticides (dichloro diphenyl trichloroethane), toxic metals (Cd, Hg, Ag, Co, Cr, Ni, Pb, Zn, and Cu), and others may cause deleterious effect in marine life. The pollutants can enter the food chain and may even reach the highest trophic levels [2]. As microorganisms are among the lowest levels of the food chains, they are the first to denounce the negative effects of pollution. Among the vast variety of microorganisms, foraminifera have shown potential for effective pollution biomonitoring, apart from many other applications [1].

These unicellular microorganisms are effective environmental indicators as they respond quickly to small environmental changes. Foraminifera are abundant and preserve the changes in their tests, making possible to study even the past environmental standing. Foraminifera are widely distributed in marine environments. In fact, they have been successful inhabitants from deep oceans to brackish water lagoons, estuaries, and even rarely in freshwater streams, lakes, and so on [3].

Foraminifera are excellent bioindicators of environmental changes resulting from both natural processes and by human interference. Most of the recent studies carried out with foraminifera are focused on the application of these organisms as environmental bioindicators. Most baseline studies arrived from natural distribution patterns like those reported in the classical literature by [4–6]. After this first step, the detection of patterns associated with natural disturbances or pollution related to human activities can be carried on [7–10]. The environmental studies using foraminifera were started by the end of the 1950s. In a study performed in Santa Monica Bay, California, Zalesny [11] stated that environmental factors such as currents activity, nutrients, salinity, characteristics of bottom sediments, and especially temperature should control the distribution patterns of the living benthic species in that bay. Since then, the number of works aiming to study the response of benthic foraminifera to environmental changing, namely to pollution, increased significantly [12].

Foraminifera have been applied to study areas with different kind of pollutants and in different kind of environments including heavily polluted harbors, such as French coast, Rio de Janeiro, Montevideo, and Eastern Sicily. Armynot du Châtelet et al. [10], Debenay et al. [13], Vilela et al. [14], and Burone et al. [15] observed that foraminifera distribution in the harbor of Port Joinville, at the Atlantic French coast, was influenced by the significant increase of pollution by metals such as Cu, Pb, and Zn and sediment texture. In the most polluted areas, they observed the increasing abundance of pollution-tolerant species. Armynot du Châtelet et al. [10] observed that foraminifera density and diversity were negatively correlated with heavy metal and PAHs in four moderately polluted harbors of the French coast.

Romano et al. [16] studying the effect of heavy pollution mainly due to Hg, PAHs, and PCBs on the foraminiferal assemblages from the Augusta harbor (Eastern Sicily, Italy) observed that the clearest response of foraminifera to environmental degradation was the increased percentages of abnormal specimens exceeding the background, the increase of pollution-tolerant species, and the reduced size of the specimens.

Furthermore, they have been used as suitable bioindicators owing to their immediate response to the environmental changes such as hydrodynamic [17, 18], salinity [19, 20], pH [21], Eh [22], heavy metals [13, 15, 18, 22, 23], hydrocarbon pollution [10, 23], and organic matter [8, 24].

The main goal of this chapter is to provide information about environmental factors that may influence the patterns of distribution of living benthic foraminifera in transitional and marine ecosystems. A semienclosed coastal lagoon (Aveiro Lagoon, Portugal), an estuarine system (São Sebastião Channel, SE Brazil), a continental shelf sector (Campos Basin, SE Brazil), and a segment of continental slope (Campos Basin, SE Brazil) are used as examples. Each area has different particularities that condition the type of living foraminifera associations that occur in them, being the first two areas highly anthropized. This work also emphasizes the importance of these organisms as environmental bioindicators and their application in biomonitoring studies.

2. Case study: Aveiro Lagoon (Portugal)

The Aveiro Lagoon is located on the Western Atlantic coast of Portugal (40°38′N, 8°45′W, **Figure 1**). It is 45 km long and 10 km wide and covers an area between 66 and 83 km^2 at low and high spring tides, respectively [25]. It is composed by a series of channels among intertidal areas such as mudflats, salt marsh, and old saltpans. It is connected to the ocean through a single artificial inlet (~350 m wide and 2 km long) fixed by two jetties (**Figure 1**). Four main channels radiate of the lagoon mouth: Mira, S. Jacinto/Ovar, Ílhavo, and Espinheiro channels (**Figure 1**), with lengths of 20, 29, 15, and 17 km, respectively. The inner lagoonal area receives the contribution of several rivers. The freshwater is supplied mainly by Vouga and Antuã rivers with average flows over 50 and 5 $m^3\ s^{-1}$, respectively [26] and in less way by small rivers such as Boco river, which flows into the south Ílhavo channel, and the Caster river, which flows into the north of Ovar channel, with an

Figure 1.
(a) Aveiro Lagoon location; (b) samples location in Aveiro Lagoon; (c) a detail of the lagoon mouth (adapted from Google Earth).

average flow less than 1 $m^3 s^{-1}$ [25]. According to Dias et al. [27], each of the main channels may be considered as presenting features of separate estuaries, owing to typical estuarine longitudinal gradients of water salinity and temperature, with values close to the characteristics of the oceanic water near the inlet and close to freshwater furthest upstream.

The average depth of the lagoon is approximately 1 m relative to the local chart datum (2.0 m below mean sea level) [27] although there are deeper areas close to the lagoon mouth, where depths may reach values of ≈30 m. Nevertheless, the bathymetry in this zone has changed over time due to both natural and anthropogenic causes [28].

The Ria de Aveiro is a tide-dominated lagoon with minimum tidal range of 0.6 m (neap tides) and maximum tidal range of 3.2 m (spring tides). Near the lagoon entrance, tidal current velocities are strong [29], but they weaken in more internal lagoon areas. Furthermore, the noticeable wind effect can have an important influence on the lagoon circulation. Particular circulation patterns mainly in shallow areas and wide channels can be induced by strong winds [25].

Regarding the environmental quality of the lagoon, in internal areas near the river mouth, there are high concentrations of metals and organic matter [18]. The Laranjo Bay area (**Figure 1**) is also polluted owing to release wastewaters of plants of the chemical complex of Estarreja [30].

2.1 Methods applied in Aveiro Lagoon

The case study of Aveiro Lagoon, commonly known as Ria de Aveiro, was based on the analysis of geochemical, sedimentological, and environmental parameters combined with living benthic foraminifera. Environmental parameters and some foraminiferal data were based on Martins et al. [18, 31, 32]. The sediments were collected, in Aveiro Lagoon channels, in 255 stations, in 2006/2007 with an adapted Petit Ponar Grab sampler (opening at both extremities), and using a ZOE I boat (**Figure 1**), they were analyzed. Water depth was measured with the boat sonar, and the stations were located with a GPS. The upper sediment layer (about 1 cm) was scraped for textural, geochemical, and microfaunal (living benthic foraminifera) analyses at each site. The sediments sampled for geochemical analysis were immediately cool preserved on board. The samples collected for foraminifera studies were kept in alcohol (90%) stained with Rose Bengal (2 g of Rose Bengal in 1000 ml of alcohol). Rose Bengal was used to differentiate between living and dead foraminifera [33]. Temperature, salinity, pH, and potential redox (Eh) were measured in water and sediments in each site.

Samples for grain size and geochemical analysis were dried to constant weight in an oven for about 72 h, at 45°C, and stored for subsequent analysis. The procedures used for sedimentological and geochemical analyses are described in detail in Martins et al. [18, 31, 32]. The description of the foraminiferal analysis can be observed in Martins et al. [18, 31, 32]. The number of species per sample (S) and the diversity index of Shannon (H′) [34] were determined. The equitability also was determined according to Pielou [35] and S is the total number of species in a sample [36].

2.2 Results obtained in Aveiro Lagoon

2.2.1 Abiotic data

Sedimentary samples were collected at water depths varying between 0.5 and 30 m. Water temperatures varied between 10.5 and 26.0°C and salinity from 6.2 to 33.7. Higher temperatures were recorded in the innermost part of the main channels and salinities near the lagoon mouth and in the channels with strong marine influence. In sediments, Eh values ranged from 134 to −222 mv, and pH between 4.2 and 10.9. The lower Eh values were found in Aveiro City canals and Murtosa channel. Some sites of Murtosa channel also have low pH values.

Sediment mean grain size (SMGS) varied between 19.7 and 3660.2 μm and fine fraction (fines; <63 μm) between 0 and 97.7%. Total organic matter (TOC) content in dry sediments ranged from 0.1 to 7.7% (**Figure 1**). Concentrations of potentially toxic elements (PTEs) varied for Zn 2-684 mg kg^{-1}; Pb 7-851 mg kg^{-1}; Cu 0.03-121 mg kg^{-1}; As 03-119 mg kg^{-1}, and Cr 78-0.03 mg kg^{-1}. The highest TOC and PTEs contents were found in protected areas.

2.2.2 Foraminiferal assemblages

Living specimens density (no per gram of sediment fraction 63–500 μm) were < 2300 n°/g. Higher densities were found in protected areas of channels with a good connection with marine waters. Ninety species of living foraminifera were found in the Aveiro Lagoon. Number of species per sample (SR) varied from 0 to 28 and Shannon index values (H) were < 2.8. The most frequent species in living foraminiferal assemblages of Aveiro Lagoon are *Ammonia tepida* (<40%) and *Haynesina germânica* (<40%), which were found in all of the sites. Other species also reach relatively high relative abundance, at least locally, such as *Elphidium margaritaceum* (<54%), *Lepidodeuterammina ochracea* (<52%), *Lobatula lobatula* (<45%), *Rotaliammina concava* (<32%), *Bolivina ordinaria* (<31%), *Cibicides ungerianus* (<19%), *Planorbulina mediterranensis* (<17%), *Cribroelphidium excavatum*, *Elphidium gerthi* (<14%), *Elphidium complanatum* (<14%), *Bolivina pseudoplicata* (<13%), *Remaneica helgolandica* (<13%), *Bulimina elongata/B. gibba* (<10%), *Elphidium williamsoni* (<6%), *Gavelinopsis praegeri* (<6%), *Trochammina inflata (<5%)*, *Elphidium crispum* (<6%), *Cribroelphidium excavatum* (<5%), *Quinqueloculina seminula* (<5%), and *Cribrostomoides jeffreysii* (<5%). Other species, such as *Buliminella elegantissima*, *Miliammina fusca*, *Haplophragmoides manilaensis*, *Entzia macrescens*, *Tiphotrocha comprimata*, *Ammoscalaria pseudospiralis*, *Arenoparrella mexicana*, *Siphotrochammina lobata*, *Ammobaculites balkwilli*, and *Eggerelloides scaber*, occur in general with percentages less than 5%.

2.3 Discussion of the results obtained in Aveiro Lagoon

The higher values of SMGS are common in samples collected along the inlet and S. Jacinto channels where the tidal currents are stronger and reach frequently velocities >2 m s^{-1} [29], and in stations of other channels due to stronger currents activity in interaction with local topographic effects. Tidal currents affect not only the sediments' texture but also their chemical composition in Aveiro Lagoon [37, 38]. The sediments are coarse-grained and have low organic matter content where the currents are strong. Under low currents activity, fine-grained sediments enriched in organic matter are accumulated. The heterotrophic activity in Aveiro Lagoon is intense [39], resulting in negative sedimentary redox potential values in many areas mostly where fine sediments and high organic matter contents are accumulated. As the region surrounding the Aveiro Lagoon is densely populated, in the most confined areas located near cities and villages or close to the rivers' mouths, higher available concentrations of PTE (such as Cr, Cu, Ni, Pb, and Zn) can be found. Highest PTE values were found, for instance, in Aveiro city and Murtosa Channel and the lowest values in the lagoon entrance.

In Aveiro Lagoon, in addition to the salinity and organic matter contents, the hydrodynamical conditions have an important influence in the pattern of distribution of benthic foraminifera assemblages. Living foraminifera density tends to increase in fine-grained sediments enriched in organic matter.

Most of the living species found in the Ria de Aveiro are typical of European estuarine environments [40, 41], of worldwide transitional environments [42], and some are present in the nearby continental shelf [43, 44]. Species such as *H. germanica*, *A. tepida*, *C. excavatum*, and *T. inflata* are typical of coastal and transitional environments [45, 46] and are quite common in Ria de Aveiro.

In Aveiro Lagoon, the agglutinated species that are known to be well adapted to a wide range of salinities [47] predominate in different ecological niches, all of them characterized by high environmental stress. *Lepidodeuterammina ochracea* and *Rotaliammina concava* dominate in very strong hydrodynamical conditions at the lagoon entrance. Instead *Miliammina fusca*, *Haplophragmoides manilaensis*, *Entzia macrescens*, *Tiphotrocha comprimata*, *Ammoscalaria pseudospiralis*, *Arenoparrella mexicana*, *Siphotrochammina lobata*, and *Ammobaculites balkwilli* reach the highest relative abundance but have low densities in low salinity waters near the rivers' mouth and in sediments with relatively low Eh and pH values, where the abundance of calcareous species decline [23, 37]. According to Fatela et al. [21], the low pH values in sediment pore water limit the episodic presence of calcareous foraminifera. Low pH levels coupled with the reactivity of biogenic carbonates may promote dissolution and destruction of calcareous tests [48].

The diversity and species richness tend to increase in the deeper areas under greater oceanic influence where there is also an increase of, for instance, *E. margaritaceum*, *L. ochracea*, *L. lobatula*, *R. concava*, *B. ordinaria*, *C. ungerianus*, *P. mediterranensis*, *E. gerthi*, *E. complanatum*, *B. pseudoplicata*, *B. elongata*/*B. gibba*, *G. praegeri*, *E. crispum*, and *C. jeffreysii*. These species seem to prefer more saline and oxygenated waters and less impacted environments and thus are named as "marine species" [43, 44].

Excess of organic matter linked with fine-grained sediments can lead to depressed levels of oxygen in the sediment pore waters, which may cause stress to benthic foraminifera [49]. However, *H. germanica, A. tepida, Bolivina ordinaria, Bolivina pseudoplicata*, *T. inflata*, and *C. excavatum,* for instance, can occur in such conditions, which means that they tolerate better the negative effects of eutrophication than, for example, *L. ochracea*, *L. lobatula*, *R. concava*, *C. ungerianus*, *P. mediterranensis*, and *G. praegeri*. However, it is known that benthic foraminifera are very tolerant to oxygen depletion, and some species appear to be resistant to hypoxic and periodic anoxic conditions [50].

According to Armynot du Châtelet et al. [51], the relative abundance of *A. tepida* is typically favored by an increase of total organic matter, meaning food resources. *Ammonia tepida* has been invariably reported as a potential bioindicator of pollution at the majority of the coastal polluted sites [1]. In general, the sites polluted with sewage rich in toxic metals had low foraminiferal abundance, high percentages of *A. tepida*, low percentage of epiphytic species, and more deformed fauna [46]. In this work, *A. tepida* is present in the most polluted sediments of the Aveiro Lagoon, but it seems to be not firstly related to the PTE enrichment. According to Armynot du Châtelet et al. [51], the relative abundance of *A. tepida* is typically favored by an increase of total organic matter, meaning food resources. A few workers, however, suggested that the preference of *A. tepida* for fine-grained organic carbon-rich sediments may be the reason for its dominance in polluted regions [1].

In Aveiro Lagoon, *H. germanica* is mostly associated with confined lagoonal sites with high content in organic matter under low currents activity and waters with relatively high salinity. This species probably displays an opportunist behavior benefiting of the organic matter supply (food) and tolerating low levels of oxygen. *Haynesina germanica* is a mid-latitudinal, temperate, and euryhaline species that populates shallow water muddy and phytal environments of salt marshes, intertidal habitats with salinities that generally range between 1 and 30, and optimal temperatures between 12 and 22°C [52]. Armynot du Châtelet et al. [10] have also shown that *H. germanica* is a successful pioneer species in polluted estuarine environments and in rich organic matter sediments and is tolerant to heavy metals. This species seems to be quite tolerant to higher concentrations of metals, namely Zn, Pb, and Cu, in Aveiro Lagoon.

The results obtained in Aveiro Lagoon also indicate that most of the species that live in this lagoon, mainly those that drive to the most internal and confined areas of the lagoon, should tolerate the stress caused by eutrophication and relatively high concentrations of PTE, namely *H. germanica*.

In general, the density and diversity of foraminifera are low in the lagoon. However, in the most impacted zones the density and diversity of foraminifera become even smaller. In fact, the increase in pollutants has been reported in general, as being marked by a decrease in species diversity with increased abundance of stress-tolerant species and high percentage of abnormalities [45, 46].

It is known that the distribution of the living assemblages is strongly affected by the estuarine dynamic, since foraminifera react within less than 1 month to changes of environmental conditions [17]. The distribution of the living foraminifera species results in several blooms throughout the year, for this reason the abundance and diversity of foraminifera is naturally temporally variable [17]. Living assemblages of foraminifera can be quite variable over time depending on the variability of the physicochemical parameters (according to weather changes). Therefore, monitoring studies may provide data not only on the response of species to the variation of environmental parameters but also on the gradients of natural and/or anthropogenic environmental impact.

3. Case study: São Sebastião Channel, SE coast of Brazil

Deciphering the impacts of domestic and industrial pollutants is difficult because they often occur together in sheltered coastal environments (bays or estuaries). When they occur separately, it is often in environments with different natural conditions, which makes comparison problematic. São Sebastião Channel (SSC) is an open area where industrial and domestic effluents are separately disposed, but under similar natural conditions, offering the opportunity to compare their impact on benthic biota [53].

SSC, located between the latitudes 23°40′S and 23°53.5′S and the longitudes 45°19′ and 45°30′W, is a 25 km stretch, which separates the continent from São Sebastião island (**Figure 2**). SSC width ranges from 2 km in its central portion and 7 km in its southern and northern ends. Its axis, where the largest depths (30–50 m) are found, is located closer to the island, due to the erosion and/or structural conditioning of the bottom. The smaller depths (6 m) occur on the continental side of the channel. The southern and northern ends have depths of 25 and 20 m, respectively (**Figure 2**).

The water circulation in the channel is characterized by alternate northerly and southerly movements, with a periodicity of days that is not directly influenced by tidal currents [54]. Geometry and topography of the channel bottom produce more intense longitudinal currents on the insular side, with speeds of up to 1.0 m s^{-1} toward the north and 0.7 m s^{-1} toward the south.

In SSC, there are some areas where the anthropic influence is quite intensified. Among these, the central region of the channel has the largest petroliferous terminals of South America, "Dutos e Terminais Centro Sul" (DTCS) of PETROBRAS. According to Duleba et al. [53], the DTCS generates two types of

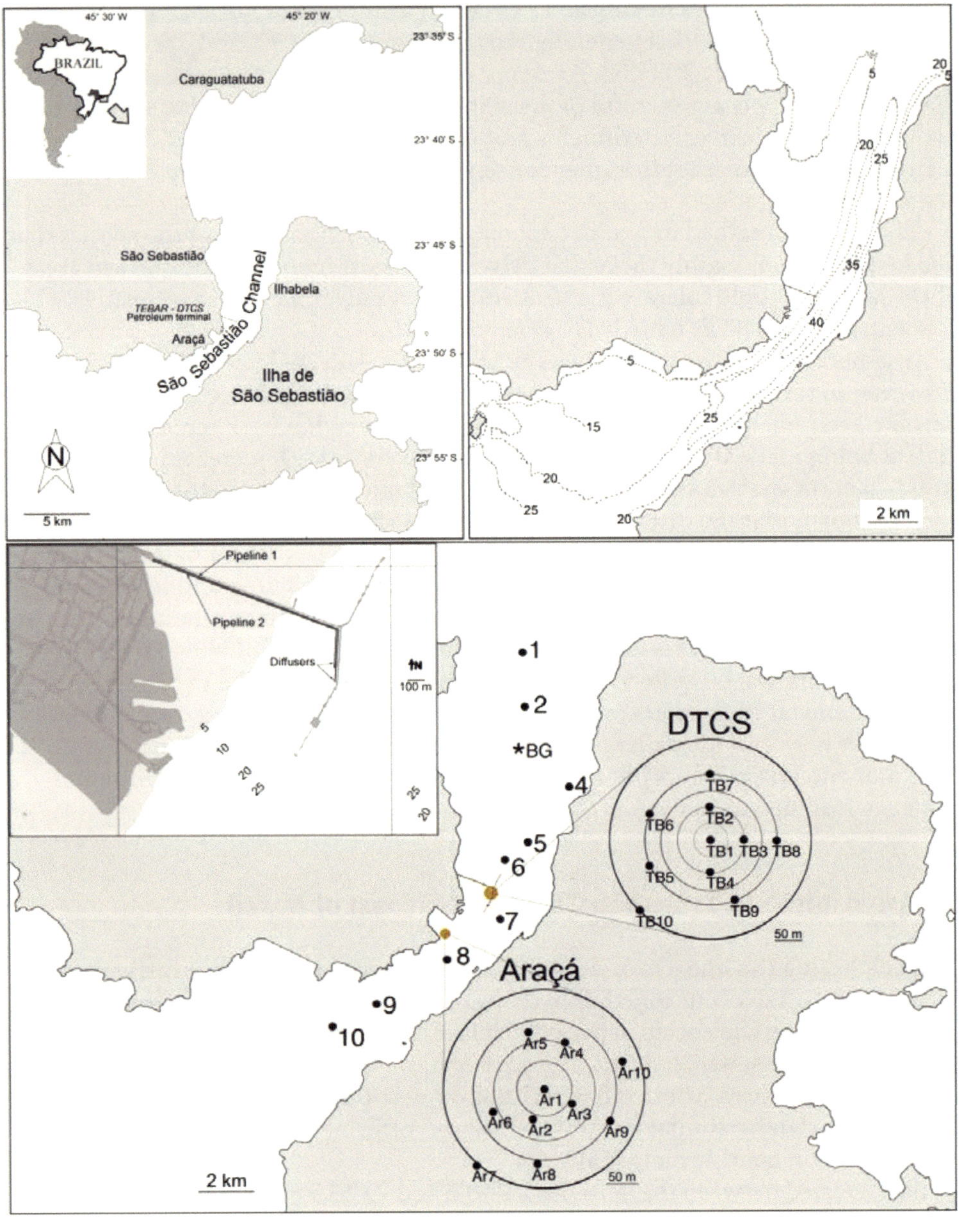

Figure 2.
Study areas and sampling grid in São Sebastião Channel (SE coast of Brazil).

liquid effluents. The first type consists mostly of water, separated from oil by density during transportation from drilling platforms in the oil tankers; the second type consists of rainwater and industrial water from the DTCS, contaminated by oil. These waters are treated in the wastewater treatment plants (ETE—Estação de Tratamento de Efluentes), where they are first separated from residual oil by adding a solution of polyelectrolyte [55]. Then, a series of treatments, using hydrogen peroxide at several pH levels, allows the oxidization of sulfides and phenol before an ultimate neutralization of the effluent. Approximately 15,000 m^3 of produced waters are treated every month. According to Fortis et al. [55], contaminated rainwater and industrial water are treated in systems that use mostly decantation to separate oil from water (SAO = Separação de Águas Oleosas). In addition, if the quality of the outflowing water does not satisfy the legal regulations, it is sent for treatment in the ETE. After treatment, effluents from both the ETE and SAO are mixed before being discharged through two submarine pipelines (with 1600 and 1400 m long), which end in diffusers, located at a depth ranging from 20 to 25 m [55].

The oil separation techniques, used in the wastewater treatment plants, primarily remove particulate matter and dispersed oil, while dissolved hydrocarbons remain in the discharged water [55]. The treated water is generally enriched with ammonia [55] and dissolved ions of sodium, potassium, magnesium, chloride, and sulfate, leading to salinity levels of up to 52.8 [55]. The treated water also has elevated levels of some heavy metals, as well as corrosion and scale inhibitors, biocides, dispersants, emulsion breakers, and other chemicals [56].

Close to DTCS, there is the submarine outfall of Araçá, which transports almost all the domestic effluents from the São Sebastião city. That emissary has a total length of 1061 m and a diameter of 400 mm. Its diffuser has a length 10.1 m and is located at Araçá point, at a depth of 8 m. The discharge speed per final discharge is 91.46 $l\ s^{-1}$ [46].

3.1 Methods applied in São Sebastião Channel

Sediment samples from DTCS (named TB) were collected near the outfall diffusers in September 2005, by the Environmental Agency of the São Paulo State, Brazil (CETESB—"Companhia Ambiental do Estado de São Paulo"). This authority evaluates the adequacy of the wastewater plants projects, the definition of environmental monitoring programs, and the regulation and enforcement of the water quality compliance. The sampling grid, consisting of 10 sampling points, was located in an area of 125,000 m^2 surrounding the diffusers. In addition, samples from 10 stations along the São Sebastião Channel (SSC) and 10 stations around the Araçá (AR) domestic sewage outfall were used for comparison. The geographical positions of the sampling stations were determined using the global positioning system (GPS), with the UTM datum SAT 69.

Surface sediment samples were collected for the following analyses: (i) grain size analyses; (ii) geochemical analyses; and (iii) determination of living benthic foraminifera. Textural, trace elements, and foraminifera data from TB, AR, and SSC areas were previously studied by Teodoro et al. [46, 53], which described in detail the methods of sedimentological and geochemical analysis.

Concerning the foraminiferal study, immediately after sampling, the samples were fixed with 70% alcohol stained with 1 g Rose Bengal, to distinguish stained (living) from unstained (dead) benthic foraminifera [57]. Aliquots of 10 cm^3 of sediment were washed through two sieves: 0.5 and 0.063 mm. The obtained fractions were dried, and the foraminifera were separated from the sediment by flotation using trichloroethylene. In samples with a low number of foraminifera, aliquots of 10 cm^3 were successively analyzed for to count of at least 100 stained individuals [33, 58]. Therefore, about 100 or more stained foraminifera were handpicked for identification and counting at each station. Foraminiferal density 1 (density 1) is expressed as the number of foraminifera per volume of sediment and density 2 is number of foraminifera per 10 cm^3 [59]. Foraminiferal assemblages structure was analyzed by using the Shannon index (H′) [34], the equitability (J′), calculated according to the Pielou index [35], and the species richness, calculated from density 1.

Canonical correspondence analysis (CCA) was used to investigate the relationship between foraminifera and sedimentological variables of the three areas: TEBAR, Araçá, and São Sebastião Channel (see details in [53]). The Monte Carlo permutation test (999 permutations) was used to assess the statistical significance of the correlations (at $p < 0.05$ and $p < 0.01$).

3.2 Results obtained in São Sebastião Channel

3.2.1 São Sebastião Channel

A total of 88 living species were identified in the São Sebastião Channel [53] belonging to the suborders of Rotaliina (63 species), Textulariina (20 species), and Miliolina (only one species). The volume of analyzed sediment needed to obtain at least 95 stained foraminifers ranged from 10 to 60 cm^3. Density 1 values ranged from 95 specimens per 60 cm^3 of sediment to 296 specimens per 10 cm^3 of sediment. Density 2 values ranged from 16 to 296 specimens per 10 cm^3 of sediment. The highest densities were identified at stations SSC3 (266 specimens) and SSC6 (296 specimens). Richness values varied from 12 to 33 species. The H′ and J′ values varied between 1.59 and 3.25 and between 0.64 and 0.93, respectively.

Ammonia tepida was the most abundant species in almost all samples (5–56.1%). The following species presented significant relative abundance: *Ammonia parkinsoniana* (2–19.6%), *Bolivina striatula* (0.8–11.6%), *Globocassidulina crassa* (<18.8%), *Globocassidulina subglobosa* (<10.2%), *Nonionella opima* (<9.1%), *Buliminella elegantissima* (2–8.5%), *Bolivina fragilis* (<8.5%), *Bulimina marginata* (<5.9%), *Pseudononion japonicum* (<5.8%), *Hopkinsina pacifica* (<5.3%), *Rosalina floridensis* (<6.1%), *Gavelinolepsis praegeri* (<5.5%), and *Hanzawaia boueana* (<5.1%). At the reference station (SSC3), *A. tepida* had the highest relative abundance (42.5%), followed by *A. parkinsoniana* (16.2%), *B. striatula* (7.9%), and *B. elegantissima* (7.5%). At this station, 32 species were recognized.

3.2.2 Dutos e Terminais Centro Sul (DTCS)

Throughout this area, 45 species were identified as belonging to Rotaliina (37 species), Textulariina (6 species), and Miliolina (2 species). Foraminiferal densities ranged from 0.5 (TB9) to 25 (TB7) specimens per 10 cm^3 of sediment. Owing to this low density, the volume of analyzed sediment needed to obtain 95 stained individuals varied from 40 to 190 cm^3. Species richness varied from 12 to 23 per 95 foraminifera. The H′ and J' values ranged from 1.5 to 2.4 and from 0.56 to 0.71, respectively. Both indices presented low values, indicating low species diversity, due to the dominance of few species.

Ammonia tepida was the most abundant species in all the samples (38.5–66%). The following species also had significant relative abundance: *Pararotalia cananeiaensis* (<20%), *B. elegantissima* (0.9–11.8%), *A. parkinsoniana* (<7.3%), *C. lobatulus* (<6.6%), *B. striatula* (<6.4%), *B. marginata* (1–6.3%), *B. ordinaria* (0.9–6%), *Bolivina compacta* (<5.5%), and *Rosalina floridensis* (<5%).

3.2.3 Araçá Outfall

In this area, 51 species [53] were identified as belonging to the suborders Rotaliina (33 species), Textulariina (11 species), and Miliolina (7 species). Foraminiferal densities ranged from 28 to 98 specimens per 10 cm^3 of sediment. A volume ranging from 10 to 40 cm^3 was analyzed to obtain 95 stained individuals. Species richness values varied from 13 to 28 species. The H′ and J' values varied from 0.70 to 2.64 and from 0.69 to 0.85, respectively.

Ammonia tepida was the most abundant species in all samples of the Araçá region, with a relative abundance ranging between 24.7 and 47.3%. The following species also had significant relative abundance: *P. cananeiaensis* (1.8–17%), *C. lobatulus* (<11.9%), *B. ordinaria* (1.1–11.3%), *R. floridensis* (<8.6%), *B. elegantissima* (<7.8%), *B. striatula* (3.2–7.7%), *P. japonicum* (<6.1%), *G. crassa* (<6.1%), and *A. parkinsoniana* (<5.4%).

3.3 Discussion of the results obtained in São Sebastião Channel

Sediments in the DTCS area were silty with high concentrations of total organic carbon (1.7–2.4%), total nitrogen (0.2–0.3%), total sulfur (0.4–0.6%), and total phosphorous (0.12–0.18%) and inorganic phosphorous (0.07–0.11%). These values were higher than those in sediments collected in the SSC and Araçá regions. The sediments' concentrations of As, Cd, Cr, Cu, Hg, Ni, Pb, and Zn in the SSC and AR regions were lower than their corresponding probable effect levels (PELs) [53]. However, sediments near the DTCS were enriched with As, Cu, and Ni, whose concentrations exceeded their corresponding threshold effect levels (TELs).

Despite potentially considerable pollution sources, mostly around DTCS, the contamination of sediment, as measured through geochemical analyses, is moderate. This probably results from the dispersion of effluents by the currents that affect the São Sebastião Channel. However, even if they are dispersed and do not accumulate within the sediment, pollutant may affect the benthos since all habitats exposed to all types of contaminants experience decreased biodiversity [60]. Indeed, the low densities of foraminifera around the DTCS diffusers illustrate the impact of environmental stress on the benthos.

In the DTCS area, it was necessary to search 50–190 cm^3 of sediment to find 100 living specimens (an average of 9 ± 6 individuals per 10 cm^3 of sediment) [53]. In the SSC and Araçá areas, a maximum of 40 cm^3 of sediment was enough to locate 100 living specimens (an average density of 62 ± 22 foraminifera per 10 cm^3 of sediment).

Organic matter may favor microfauna [7], or it may be responsible for decreasing microfauna density and richness [10, 15]. The toxic threshold depends on the nature of the organic matter and its concentration in the sediments [10]. The degradation of organic matter requires large quantities of oxygen; thus, when the flux of organic matter exceeds the degradation rate, a benthic hypoxia or even anoxia can occur. In this sense, the microfauna is compelled to change: stenobiotics can disappear and an abundance of tolerant species may be observed [10, 15, 31].

Pararotalia cananeiaensis is an herbivorous, epifaunal species characteristic of a marine environment. It is abundant in dead assemblages all over the SSC. The positive correlation of *P. cananeiaensis*, together with *D. floridana*, with TP and Cd appears to be an indication that they are tolerant to these elements and to the associated contaminants. However, the ecological preference of *P. cananeiaensis* is not yet well known in Brazil (Debenay et al., 2001c)***. Near the Araçá domestic submarine outfall, Teodoro et al. [46] reported *P. cananeiaensis* living preferentially at stations with high sulfur content ($r = 0.86; p < 0.001$), particulate organic matter ($r = 0.62; p < 0.001$), and silt ($r = 0.75; p < 0.01$). In the present study, no relationship between abiotic parameters and relative abundance of *P. cananeiaensis* was recorded.

The species related to reducing muddy sediment, with a moderate concentration in Cr and a noticeable content of organic matter, were *B. marginata*, *B. elegantissima*, *B. compacta*, *A. tepida*, and *A. parkinsoniana*. Most of these species are recognized in the literature as tolerant to high organic matter flux and as able to survive in low oxic conditions. Such is the case of *B. marginata* [61] and *B. elegantissima* [14]. Bandy et al. [62] noted that *B. elegantissima* and *B. marginata* tend to be abundant in areas affected by pollutants. In this study, the highest abundance of bolivinids and buliminids was observed in stations positioned in the central part of the channel: stations SSC7, SSC8, and SSC9.

Ammonia tepida is a eurybiotic species characteristic of near-shore areas and paralic environments [63]. The tolerance of *A. tepida* to adverse conditions, including organic and chemical pollution, has long been reported in both field studies and culture studies [10]. Its potential application for pollution monitoring is well established. It reached the highest relative abundance in DTCS the polluted samples and, to a lesser extent, in Araçá. This higher proportion of *A. tepida* is due to a decline of stenobiotic species. The higher abundance of tolerant species in these areas indicates that the benthos is significantly impacted by both organic (Araçá) and chemical (DTCS) pollution and suffers with a greater impact of chemical pollution.

4. Case study: continental shelf of the Campos Basin (SE, Brazil)

The continental shelf of Campos Basin is located on the southeast coast of Brazil between latitude 21° and 23°S, the southwest margin of the Atlantic Ocean, which comprises part of the area occupied by the Marine Sedimentary Campos Basin between the 25 and 150 m isobaths (**Figure 3**). The sedimentary contribution to this region is restricted to the rivers Itapemirim, Paraíba do Sul, Macaé, and São João, and most of the material derived from rivers and coastal erosion seems to be retained in coastal waters; what exceeds in this region is readily carried by oceanic currents. Detrital materials derived from the river discharge of the Paraíba do Sul River are distributed throughout the inner shelf by developing small muddy zones and large mud accumulations in an area adjacent to Búzios and Cabo Frio, where the currents have energy close to zero, 150 km south to the mouth of the river [64].

The South Atlantic Central Waters (SACW), a colder, nutrient-rich water mass, enters the continental shelf increasing local primary production and associated secondary productivity. Many studies have been carried out to investigate aspects related to the coastal upwelling of Cabo Frio, south of the Campos Basin [65].

4.1 Methods applied in Continental Shelf of the Campos Basin

Sampling methods, preparation of samples, control of quality, and methods of data analysis are described in detail in [66]. Living benthic foraminifera of the Campos Basin continental shelf were studied in 239 samples collected in nine transects (A–I) perpendicular to the coast line. The sampling was performed during the dry season of 2008 and the rainy season of 2009. The dry season, in this region, occurs in winter and corresponds to lowest precipitation and less frequent upwelling events. By contrast, the rainy season takes place in summer and corresponds to the period with higher precipitation and with more frequent upwelling events.

At each transect, five isobaths (25, 50, 75, 100, and 150 m) were sampled with a very large and modified (with an upper opening) Van Veen, which functions as a box corer. In each isobath, three independent samples successfully taken with a 10 cm × 10 cm × 2 cm, a "quadrat" yields samples with 200 cm^3. A fixative solution (4% formalin buffered with sodium borate) with Rose Bengal stain was immediately added to fix and evidence the protoplasm of living foraminifera. In the laboratory, 20 cm^3 of sediment from each sample was separated for analysis of living foraminifera.

The foraminiferal samples were washed trough a 63 μm sieve, dried in an oven (<60°C), and then picked under a stereomicroscope. Density values are equivalent to the total number of living individuals in 20 cm^3 (volume) or 10 cm^2 (area). Although the 63 μm mesh size was used as the size limit in the washing and screening of samples, many individuals smaller than 63 μm were found alive adhered to the grains; they were removed and incorporated into the slides for study. Careful quality control ensured similar patterns of screening and identification of living foraminifera, making the differences between pickers minimized. The species identification was based on [67, 68], and other specific references, as well as by the analysis of museum collections. The biomass was calculated by the volumetric method [33, 69].

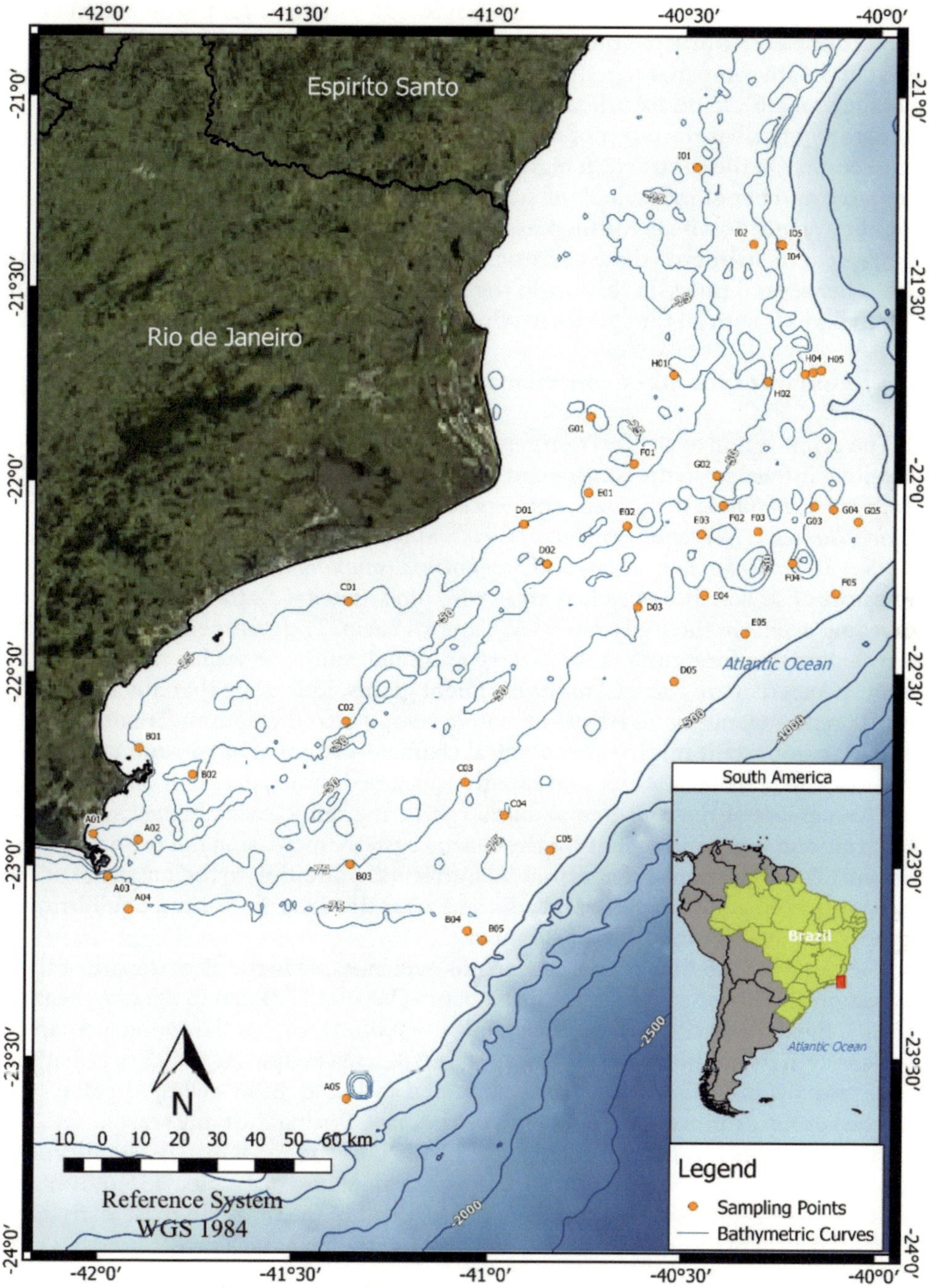

Figure 3.
Continental shelf of Campos Basin area. The transects A–I and their sampling points 1–5, according to the isobaths (25 m = 1; 50 m = 2; 75 m = 3; 100 m = 4; and 150 m = 5).

According to Pianka [70], k-strategist or conservative species have greater body size, longer life cycle, and population size largely constant in time, being close to the capacity of support of the environment; r-strategist species are known for their opportunistic behavior, small size, short life cycle, and very variable population size without adjustment balance in relation to available resources (mainly space and food). R-strategists can proliferate opportunistically and vary considerably its absolute abundance. According to these characteristics, Warwick [71] developed a method based on the abundance and biomass curves (ABC) for the detection of perturbation and to analyze the species response to environmental changes. The ABC curves are compared together with the W statistic calculation [71]. According to Magurran [36], in undisturbed environments, one or two species dominate the biomass and this has the effect of raising the biomass curve in relation to the abundance curve. On the contrary, in highly disturbed environments, a few species have very large number of individuals of small size, but since these species have small size, they do not dominate the biomass and so, the abundance curve is above the biomass curve. In intermediate conditions, the biomass and abundance curves cross each other several times. In this work, the ABC curves were generated for each group in both periods with all the species in which it was possible to calculate the biomass.

4.2 Discussion of results of continental shelf of the Campos Basin

The composition of the foraminifera assemblages allowed to detect areas under upwelling influence on the continental shelf. *Epistominella exigua*, *Adercotryma glomeratum*, *Bulimina marginata*, *Pappina compressa*, *Angulogerina angulosa* s.l., *Nonionella stella*, *Nonionella opima*, *Hopkinsina pacifica, Bolivina fragilis, Bolivinellina translucens, Fursenkoina pontoni*, and *Stainforthia complanata* can be considered as indicators of seasonality-enriched areas with phytodetrital material signaling the upwelling events in the study region of Campos Basin. The continental shelf presents many encrusting foraminifera, both large and small adults, as well as juveniles of larger species that are adhered to the sediment grains, indicating that it is a region with a predominance of high hydrodynamic energy near the bottom. Small individuals play an important role in the ecological characterization of oligotrophic areas where organic carbon is scarce and rapidly consumed, transported, or oxidized.

The increase of the continental discharge during rainy season brings more inorganic and organic nutrients to the coastal system, increasing the primary production. The high supply of continental nutrients in addition to the nutrients of the coastal upwelling of Cabo Frio and adjacent areas disturbs the natural equilibrium of the biological communities of this area.

The results of the ABC curves by isobaths evidence that in the 25 m isobath of the dry season of 2008, there was some disturbance ($W = 0.0332$), but in the rainy season of 2009, the disturbance was much high ($W = -0.0071$), so that the abundance curve overlaps with that of biomass (**Figure 4**). The ABC curves indicate the 25 m isobath as the most disturbed one within the continental shelf and the upwelling/organic enriched group as an area of moderated disturbance. During the rainy season, the 25 m isobath is disturbed by natural eutrophication phenomena and it may be significantly amplified by anthropogenic activities in the catchment area of Paraiba do Sul river and other anthropogenic disturbances, once this segment of the Brazilian coast is strongly influenced by agricultural, industrial, and urban activities as well as by fishery and many other coastal activities. In the southern region, the disturbed patterns seem to be strongly related to natural phenomena due to the presence of many species who indicate upwelling events, which are well known near the Cabo Frio area.

In his final remarks, Warwick [71] does not believe that this method can be applied to meiobenthic species, because according to him, there are no obvious size differences between k species and meiobenthic r-strategists. He mentioned Oncholaimidae nematodes and Tisbidae copepods as dominant species in polluted environments and reminds that they are often larger. However, among the foraminifers, there are species known as k-strategists that correspond to the characteristics mentioned by Pianka [70], and therefore, the application of the ABC curves to assess environmental disturbance could be successful. Although the foraminifera belong to both micro- and meiofauna, the results obtained with the ABC curves in

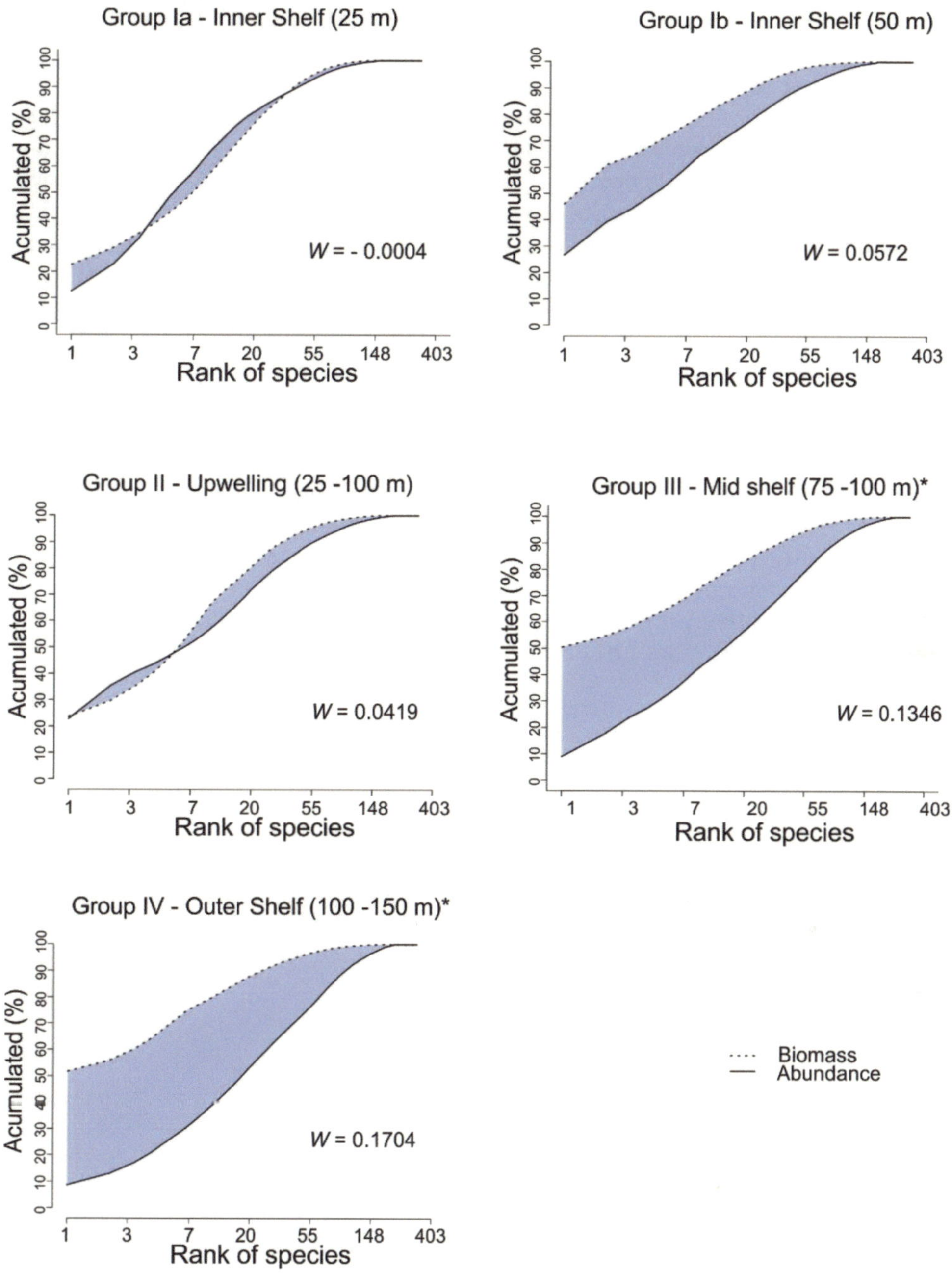

Figure 4.
ABC curves of the continental shelf assemblages of foraminifers. Groups 1–4 of the cluster analysis from rainy period (2009).

this work evidence they can be used to trace environmental disturbance. In fact, it is expected that the ABC curves can be used to identify both natural and anthropogenic stressors. However, without complementary studies that allow to qualify and quantify the disturbance origin by applying the ABC curves, we will have only the sign of the environmental disorder. On the other hand, we still have little information on the response of some species to natural and anthropogenic stressors, but this integrated study will allow the construction of a more effective monitoring plan, which can provide subsidies for preserving the integrity of the environment.

5. Case study: continental slope of the Campos Basin (SE, Brazil)

The Campos Basin is located between latitudes 21°S and 23°S in the southwestern margin of the South Atlantic Ocean (**Figure 5**). It ranges from the coastal plain to the São Paulo Plateau, the outer boundary of the salt tectonic coinciding with the boundary between continental and oceanic crusts [72]. The continental slope displays a convex profile to the north and a concave one to the south, and it is cut by several canyons, such as the São Tomé, Itapemirim, Grussaí, and Almirante Câmara [72]. The Campos Basin is considered a meso-oligotrophic system with oxygenated bottom water [73].

In Campos Basin, the oil exploration activities began in 1976, with the first maritime drilling occurring at 100 m of water depth. In 1999, exploration and production of hydrocarbons in deep waters (below 2000 m of depth) started in the Campos Basin. About 65% of hydrocarbon exploration and production activities are concentrated in marine areas deeper than 400 m [74].

The first initiative to evaluate the environmental conditions of the Campos Basin occurred in the end of 1980s, through a partnership between the Fundação de Estudos e Pesquisas Aquáticas (FUNDESPA), University of São Paulo, and Petróleo Brasileiro S/A (PETROBRAS). Regrettably, the data obtained were not widely disseminated through scientific publications [74].

From this on, other projects were executed in the Campos Basin, such as "Campos Basin Deep-sea Environmental Program" (Oceanprof) and

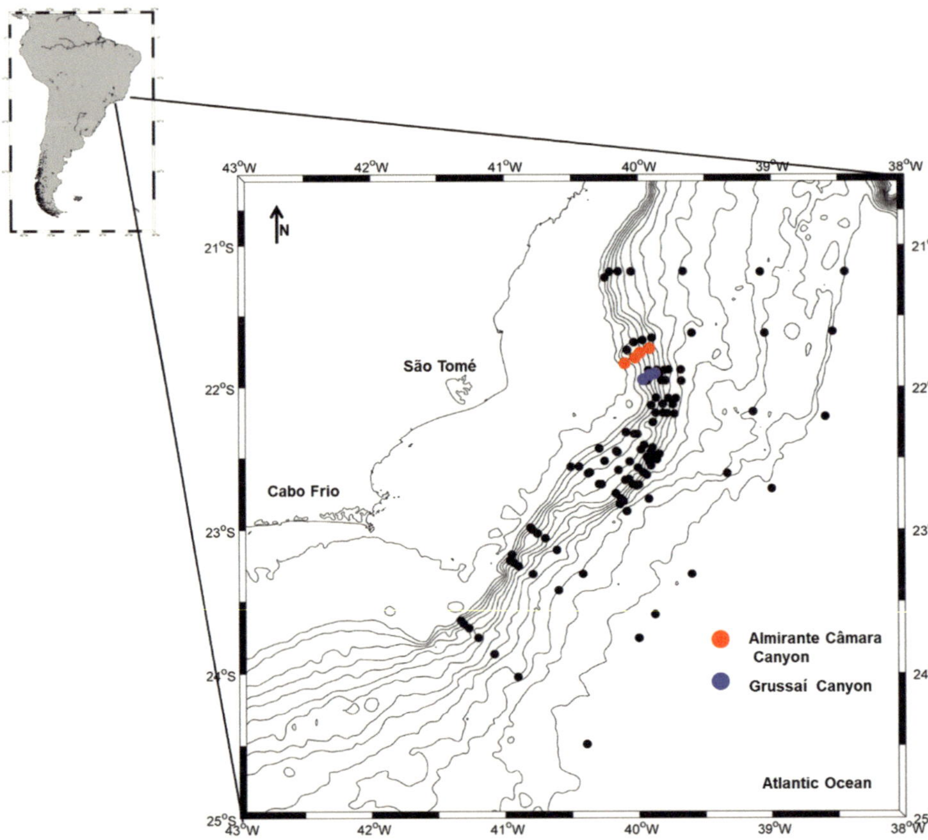

Figure 5.
Location of Campos Basin in the Southwest Atlantic Ocean and sampling sites of the study area (Oceanprof and Habitats project): red circle (station in the Almirante Câmara canyon); and blue circle (station in the Almirante Câmara canyon).

"Environmental Heterogeneity of the Campos Basin" (Habitats) projects. These projects had a multidisciplinary goal: geology and meteoceanography; organic and inorganic components of water and sediment; and the distribution and composition of biota [73, 74].

5.1 Methods applied in continental slope of the Campos Basin (SE, Brazil)

In the Oceanprof project, 41 surface sediment samples were collected by using a box corer, along transects ranging from 750 to 1950 m of water depth on the continental slope (**Figure 5**), in the austral winter of 2003. The core top (0–2 cm interval) sediment sampled in each site was used to understand the living and dead benthic foraminifera distribution patterns and ecological preferences. For this purpose, identification and quantitative foraminifera analysis were performed using the 63 μm size fraction. The variables considered for the statistical analysis included percentages of sand and mud, calcium carbonate and total organic carbon contents in the sediment, and total phosphate in the water (see details in [73]).

In the same project (Oceanprof), 20 surface sediment samples were chosen along 1050 and 1950 m depth (**Figure 5**). In this case, only living (rose Bengal stained) benthic foraminifera were analyzed in combined samples collected in four different slices of the cores (0–2, 2–5, 5–10, and 10–15 cm) to understand the distribution patterns and their ecology. In this study, the variables considered for statistical analysis included the sand and mud contents, particulate organic matter flux to the sea floor, bottom water dissolved oxygen concentrations, calcium carbonate, total organic carbon, total nitrogen, and total lipid contents in the sediment (see details in [75]).

In the Habitats project, an ecological study of living (rose Bengal stained) benthic foraminifera was performed in samples collected with a box corer, on the continental slope, Plateau of São Paulo, and canyons, during two campaigns (austral winter of 2008 and summer of 2009). The stations followed nine transects from 400 to 3000 m deep (**Figure 5**). In the Grussaí and Almirante Câmara canyons, the stations were located in four isobaths (400, 700, 1000, and 1300 m deep) and the obtained data were compared with adjacent transects on the open slope (**Figure 5**). Changes in the density, diversity, and composition of benthic foraminifera were analyzed in response to environmental factors (i.e., sand and mud contents, calcium carbonate, total organic carbon and chlorophyll-*a* concentrations, and phytopigment concentrations in the sediment) (see details in [76]).

5.2 Results obtained in continental slope of the Campos Basin and discussion

The middle slope is characterized by the dominance of different species of the genus *Bolivina*, *Cassidulina laevigata*, and *Globocassidulina subglobosa*. The occurrence of these species in association with *Cibicidoides kullenbergi*, *Epistominella exigua*, and *Uvigerina proboscidea* seems to be related to seasonal organic matter fluxes, relatively oxic bottom waters, strong bottom currents, and sandy sediments. The lower slope is inhabited by a microfauna with different characteristics, preferentially composed of epifaunal or shallow infaunal deposit feeding species (e.g., *Bolivina* spp., *Eponides weddellensis,* and *Lenticulina cultrata*) and suspension feeders that are adapted to oligotrophic conditions and high dissolved oxygen levels in the bottom waters, for example, *Rhabdammina* spp., *Rhizammina* sp. [73].

Yamashita et al. [75] concluded that besides the sediment grain size, the vertical flux of particulate organic matter seems to be the main factor controlling the spatial distribution of benthic foraminifera species in the slope of Campos Basin. The middle slope (1050 m of water depth) was characterized by relatively high foraminiferal density and a predominance of phytodetritus-feeding foraminifera such as *Epistominella exigua* and *Globocassidulina subglobosa*. The occurrence of these species seems to reflect the Brazil Current System (BCS). The abovementioned currents are associated with the relatively high vertical flux of particulate organic matter and the prevalence of sandy sediments, respectively. The lower slope (between 1350 and 1950 m of water depth) was marked by low foraminiferal density and assemblages composed of *Bolivina* spp. and *Brizalina* spp., with low particulate organic matter flux values, muddy sediments, and more refractory organic matter. The distribution of this group seems to be related to episodic fluxes of food particles to the seafloor, which are influenced by the BCS at the surface and are deposited under low deep current activity (Intermediate Western Boundary Current; [75]).

According to Sousa et al. [76], the availability and quality of the food, the energy state (stability) at the benthic/pelagic boundary, and the grain size of the substrate seem to be the most important environmental factors determining the distribution pattern of the benthic foraminiferal assemblages in the Campos Basin slope. The highest values of density, diversity, and richness, as well as the predominance of hyaline calcareous foraminifera and infaunal species, reflect a higher contribution of food received continuously in the shallower stations (400 m depth). At 700–1000 m of water depth, the density of foraminifera decreased and there was a larger presence of opportunistic species, possibly reflecting the pulse of phytodetritus. The considerable increase in the agglutinated foraminifera, the continuous decrease in the density of foraminifera, and also decrease in the values of the Benthic Foraminifera High Productivity Index (BFHP [61]) as depth increases indicate typical oligotrophic conditions in this sector of the Campos Basin (lower slope and São Paulo Plateau).

The comparison of density and species composition data in the austral winter of 2008 and austral summer of 2009 periods allows us to infer that during the winter, the food input was higher. The values of density and biomass of living benthic foraminifera allow us to suggest that the Almirante Câmara canyon is a greater entrapment site of organic matter between 400 and 1000 m isobaths in comparison to the open areas at the same isobaths [76].

Even though the stations analyzed were the same or spatially very close to each other, the response of benthic foraminifera of the Campos Basin seems to be slightly different due to the seasonality. Because of these dissimilar times, the local hydrodynamics can change and, consequently, variations can occur in the particulate organic matter flux, and quantity and quality of the organic matter on the sea floor [73, 75, 76]. The benthic foraminifera and the geochemical results of these projects showed that, despite the intense oil exploration and production activities in the Campos Basin by PETROBRAS, these specific study areas (**Figure 5**) did not suffer anthropic impact in terms of pollutants [77].

6. Conclusion

This work presents results of living foraminifera used to analyze these meiofaunal organism responses to different types of environmental disturbance in different transitional and marine settings: a semienclosed coastal lagoon (Aveiro Lagoon), an estuarine system (São Sebastião Channel), a continental shelf (Campos Basin), and continental slope environments (Campos Basin). Each area has different particularities, conditioning the type of living foraminifera associations that inhabit them, being the first two areas highly anthropized.

The dynamics of tidal currents, in interaction with the configuration of channels and local topography, generates different sedimentary facies in the Aveiro Lagoon, influencing the abundance of living benthic foraminifera. Salinity is a key factor for governing the structure (diversity) of foraminifera assemblages, as well as the concentrations of PTE, whereas the sensitive species ("marine species") avoid the inner lagoonal environments. Species such as *H. germanica*, *A. tepida*, *T. inflata*, and *C. excavatum* increase their frequencies in the most confined places and are impacted by high concentrations of PTE.

In São Sebastião Channel, the sediments near the "Dutos e Terminais Centro Sul" (DTCS) of PETROBRAS were enriched by As, Cu, and Ni, with concentrations exceeding TEL; these levels are associated with adverse biological effects. Comparatively, foraminiferal parameters (density and diversity) at the DTCS were lower than those observed in neighboring areas, even near the Araçá submarine outfall less than 3 km away. These findings lead us to conclude that wastewater treatment in DTCS is not effective in removing some chemical elements from petrochemical waste liquid. Moreover, it may negatively impact benthic fauna around the DTCS.

The distribution pattern of the living foraminiferal assemblages in the continental shelf of the Campos Basin changes depending on the bathymetry, sediment characteristics, and the supply of organic matter. The abundance of benthic foraminifera populations is strongly enlarged by the seasonal supply of phytobenthos and phytoplankton. The application of the ABC curves method in foraminiferal assemblages is a promising alternative to evaluate the environmental conditions and to access specific areas over time, and thus, they can be applied in environmental monitoring studies.

The benthic foraminifera species that occur in the deep marine system of the Southwestern Atlantic (continental slope of the Campos Basin) are mainly controlled by local hydrodynamics, which mainly controls changes in the particulate organic matter flux, quantity, and quality of the organic matter in the seafloor. It should be also considered that these parameters are affected by seasonality.

The data presented here show the importance of understanding the ecology of the benthic foraminifera species for environmental assessment of the ecosystems, and therefore for the establishment of biomonitoring procedures.

Acknowledgements

The authors would like to thank the collaboration of the Biology Department of Aveiro University, as well as Prof. Fernando Rocha of Aveiro University and Prof. Maria Antonieta Rodrigues, Universidade do Estado do Rio de Janeiro—UERJ for the support. The research conducted at the SSC was funded by Fundação de Amparo à Pesquisa do Estado de São Paulo 09/51031-8 (WD), 02/02611-2 (WD), and CETESB. We are also grateful to PETROBRAS for providing sediment samples and financial support under "Environmental Heterogeneity of the Campos Basin" and "Campos Basin Deep-sea Environmental Program" project especially to Ana Falcão and Helena Lavrado. The parameters for the organic carbon and total nitrogen were kindly provided by Dr. Carlos E., Rezende (Universidade Estadual Norte Fluminense), the particulate organic matter vertical flux by MScs. Thaisa Vicente (Fundação para Desenvolvimento Tecnológico da Engenharia), the lipids biomarkers by Dr. Renato Carreira (Pontíficia Universidade Católica do Rio de Janeiro), the bacterial biomass by Dr. Rodolfo Paranhos (Federal University of Rio de Janeiro), and the parameters for the water column by Dr. Ilson Silveira (Oceanographic Institute, São Paulo University). The authors are indebted to MScs. Nancy Taniguchi, Dr. Sueli Godoi (Oceanographic Institute, São Paulo University), Carlos E. L. Elmadjian (Institute of Mathematics and Statistics, University of São Paulo), MScs. Naira Yamamoto, Dr. Raquel Fernanda Passos, Dr. Renata Hanae Nagai (Center for Marine Studies, Federal University of Paraná), Carla Ito, Dr. Leticia Burone (Facultad de Ciencias, Universidad de la

República), Dr. Marina Fukumoto, MScs. Rodrigo Aluizio (Federal University of Paraná), Dr. Michel Michaelovitch de Mahiques (Oceanographic Institute, São Paulo University), and Dr. Aurea Ciotti (Biology Marine Center, University of São Paulo) who are gratefully acknowledged for the technical support provided. The authors are indebted Andreia Cristiane Teodoro (BTX Geologia e Meio Ambiente), Silas Gubitoso (BTX Geologia e Meio Ambiente), Silvio Miranda Prada (UNIFEO), José Eduardo Bevilacqua (CETESB).

Conflict of interest

There are no "conflicts of interest."

Author details

Maria Virginia Alves Martins[1,2]*, Cintia Yamashita[3], Silvia Helena de Mello e Sousa[3], Eduardo Apostolos Machado Koutsoukos[4], Sibelle Trevisan Disaró[5], Jean-Pierre Debenay[6] and Wânia Duleba[7]

1 Departamento de Estratigrafia e Paleontologia, Faculdade de Geologia, Universidade do Estado do Rio de Janeiro, Rio de Janeiro, Brazil

2 Departamento Geociências, GeoBioTec, Universidade de Aveiro, Aveiro, Portugal

3 Departamento de Oceanografia Física, Química e Geológica do Instituto Oceanográfico, Universidade de São Paulo, São Paulo, Brazil

4 Institut für Geowissenschaften, Universität Heidelberg, Heidelberg, Germany

5 Laboratório de Foraminíferos e Micropaleontologia Ambiental (LaFMA), Universidade Federal do Paraná (UFPR), Brazil

6 UMR 7159, IPSL/LOCEAN, Centre IRD France Nord, Bondy Cedex, France

7 Escola de Artes Ciências e Humanidades, Universidade de São Paulo, São Paulo, Brazil

*Address all correspondence to: virginia.martins@ua.pt

IntechOpen

© 2019 The Author(s). Licensee IntechOpen. This chapter is distributed under the terms of the Creative Commons Attribution License (http://creativecommons.org/licenses/by/3.0), which permits unrestricted use, distribution, and reproduction in any medium, provided the original work is properly cited.

References

[1] Suokhrie T, Saraswat R, Nigam R. Foraminifera as bio-indicators of pollution: A review of research over the last decade. In: Kathal PK, Nigam R, Talib A, editors. Micropaleontology and its Applications, India: Scientific Publishers; 2017. 265-284 pp

[2] Binelli A, Provini A. POPs in edible clams from different Italian and European markets and possible human health risk. Marine Pollution Bulletin. 2003;**46**:879-886. DOI: 10.1016/S0025-326X(03) 00043-2

[3] Solai A, Suresh Gandhi M, Rajeshwara Rao N. Recent benthic foraminifera and their distribution between Tuticorin and Tiruchendur, Gulf of Mannar, south-east coast of India. Arabian Journal of Geosciences. 2012;**6**(7):2409-2417. DOI: 10.1007/s12517-011-0514-1

[4] Murray JW. Distribution and Ecology of Living Benthic Foraminiferids. London: Heinemann Educational Books Ltda; 1973. 274 p

[5] Nichols MM. Foraminifera in estuarine classification. In: Odum HT, Copeland BJ, McMahan EA, editors. Coastal Ecological Systems of the United States—A Source Book for Estuarine Planning. Vol. I, cap. II. Charleston: The Conservation Foundation Washington, D.C. in Cooperation with National Oceanic and Atmospheric Administration Office of Coastal Environment Vff CM US, Department of Commerce KOAA Coastal Services Center. 1974. pp. 85-103

[6] Phleger FB. Ecology and distribution of recent Foraminifera. 2nd ed. Baltimore: John Hopkins Press; 1960. 297 p

[7] Nagy J, Alve E. Temporal changes in foraminiferal faunas and impact of pollution in Sandebukta, Oslo Fjord. Marine Micropaleontology. 1987;**12**:109-128

[8] Alve E. Benthic foraminiferal responses to estuarine pollution: A review. Journal of Foraminiferal Research. 1995;**25**:190-203

[9] Cearreta A, Irabien MJ, Leorri E, Yusta I, Croudace IW, Cundy AB. Recent anthropogenic impacts on the Bilbao Estuary, Northern Spain: Geochemical and microfaunal evidence. Estuarine, Coastal and Shelf Science. 2000;**50**(4):571-592

[10] Armynot du Châtelet E, Debenay J-P, Soulard R. Foraminiferal proxies for pollution monitoring in moderately polluted harbors. Environmental Pollution. 2004;**127**:27-40. DOI: 10.1016/S0269-7491(03)00256-2

[11] Zalesny ER. Foraminiferal ecology of Santa Monica Bay, California. Micropaleontology. 1959;**5**:101-126

[12] Nigam R, Saraswat R, Panchang R. Application of foraminifers in ecotoxicology: Retrospect, perspect and prospect. Environment International. 2006;**32**:273-283

[13] Debenay JP, Tsakiridis E, Soulard R, Grossel H. Factors determining the distribution of foraminiferal assemblages in Port Joinville Harbor (Ile d'Yeu, France): The influence of pollution. Marine Micropaleontology. 2001;**43**:75-118. DOI: 10.1016/S0377-8398(01)00023-8

[14] Vilela CG, Batista SD, Baptista-Neto JA, Crapez M, McAllister JJ. Benthic foraminifera distribution in high polluted sediments from Niterói Harbor (Guanabara Bay), Rio de Janeiro, Brazil. Anais da Academia Brasileira de Ciências. 2004;**76**:161-171

[15] Burone L, Venturini N, Sprechmann P, Valente P, Muniz P. Foraminiferal responses to polluted sediments in the Montevideo coastal zone, Uruguay.

Marine Pollution Bulletin. 2006;**52**:61-73. DOI: 10.1016/j.marpolbul.2005.08.007

[16] Romano E, Bergamin L, Finoia MG, Celia Magno M, Ausili A, Gabellini M. The effects of human impact on benthic foraminifera in the Augusta harbour (Sicily, Italy). In: Dahl E, Moksness E, Støttrup J, editors. Integrated Coastal Zone Management. Chichester, UK: Blackwell Publishing Ltd.; 2009. pp. 97-115

[17] Debenay J-P, Bicchi E, Goubert E, Armynot du Châtelet E. Spatio-temporal distribution of benthic foraminifera in relation to estuarine dynamics (Vie estuary, Vende'e, W France). Estuarine, Coastal and Shelf Science. 2006;**67**: 181-197. DOI: 10.1016/j.ecss.2005.11.014

[18] Martins V, Yamashita C, Sousa SHM, Martins P, Laut LLM, Figueira RCL, et al. The response of benthic foraminifera to pollution and environmental stress in Ria de Aveiro (N Portugal). Journal of Iberian Geology. 2011;**37**(2):231-246. DOI: 10.5209/rev_JIGE.2011.v37.n2.10

[19] Debenay JP, Geslin E, Eichler BB, Duleba W, Sylvestre F, Eichler P. Foraminiferal assemblages in a hypersaline lagoon Araruama (RJ) Brazil. Journal of Foraminiferal Research. 2001;**31**:133-151

[20] Martins MVA, Pinto AFS, Frontalini F, et al. Can benthic foraminifera be used as bio-indicators of pollution in areas with a wide range of physicochemical variability? Estuarine, Coastal and Shelf Science. 2016;**182**:211-225. DOI: 10.1016/j.ecss.2016.10.011

[21] Fatela F, Moreno J, Moreno F, Araújo MF, Valente T, Antunes C, et al. Environmental constraints of foraminiferal assemblages distribution across a brackish tidal marsh (Caminha, NW Portugal). Marine Micropaleontology. 2009;**70**:70-88. DOI: 10.1016/j.marmicro.2008.11.001

[22] Bernhard JM, Sen Gupta BK. Foraminifera of oxygen-depleted environments. In: Sen Gupta BK, editor. Modern Foraminifera: Kluwer Academic Publishers; 1999. pp. 201-216

[23] Vilela C, Silva Batista D, Baptista Neto JA, Ghiselli RO Jr. Benthic foraminifera distribution in a tourist lagoon in Rio de Janeiro, Brazil: A response to anthropogenic impacts. Marine Pollution Bulletin. 2011;**62**:2055-2074. DOI: 10.1016/j.marpolbul.2011.07.023

[24] Hyams-Kaphzan O, Almogi-Labin A, Benjamini C, Herut B. Natural oligotrophy vs. pollution-induced eutrophy on the SE Mediterranean shallow shelf (Israel): Environmental parameters and benthic foraminifera. Marine Pollution Bulletin. 2009;**58**:1888-1902. DOI: 10.1016/j.marpolbul.2009.07.010

[25] Dias JM, Lopes JF, Dekeyser I. Lagrangian transport of Particles in Ria de Aveiro Lagoon, Portugal. Physics and Chemistry of the Earth, Part B. 2001;**26**(9):721-727. PII: S1464-1909(01)00076-4

[26] Moreira HM, Queiroga H, Machado MM, Cunha MR. Environmental gradients in a southern estuarine system: Ria de Aveiro, Portugal, implication for soft bottom macrofauna colonisation. Netherlands Journal of Aquatic Ecology. 1993;**27**(2-4):465-482. DOI: 10.1007/BF02334807

[27] Dias JM, Lopes JF, Dekeyser I. Hydrological characterization of Ria de Aveiro lagoon, Portugal, in early summer. Oceanologica Acta. 1999;**22**:473-485. DOI: 10.1007/s10236-003-0048-5

[28] Plecha S, Silva PA, Vaz N, Bertin X, Oliveira A, Fortunato AB, et al. Sensitivity analysis of a morphodynamic modelling system applied to a coastal lagoon inlet. Ocean Dynamics.

2010;**60**:275-284. DOI: 10.1007/s10236-010-0267-5

[29] Vaz N, Dias JM, Leitão PC. Three-dimensional modelling of a tidal channel: The Espinheiro Channel (Portugal). Continental Shelf Research. 2009;**29**:29-41. DOI: 10.1016/j.csr.2007.12.005

[30] Pereira ME, Lillebø AI, Pato P, Válega M, Coelho JP, Lopes CB, et al. Mercury pollution in Ria de Aveiro (Portugal): A review of the system assessment. Environmental Monitoring and Assessment. 2009;**155**:39-49. DOI: 10.1007/s10661-008-0416-1

[31] Martins V, Silva EF, Sequeira C, Rocha F, Duarte AC. Evaluation of the ecological effects of heavy metals on the assemblages of benthic foraminifera of the canals of Aveiro (Portugal). Estuarine Coastal Shelf Science. 2010;**87**:293-304. DOI: 10.1016/j.ecss.2010.01.011

[32] Martins VA, Frontalini F, Tramonte KM, et al. Assessment of the health quality of Ria de Aveiro (Portugal): Heavy metals and benthic foraminifera. Marine Pollution Bulletin. 2013;**70**:18-33. DOI: 10.1016/j.marpolbul.2013.02.003

[33] Murray JW. Ecology and Paleoecology of Benthic Foraminifera. New York: Longman, Wiley, Harlow, Scientific, Technical; 1991. 397 p

[34] Shannon CE. A mathematical theory of communication. Bell System Technical Journal. 1948;**27**:379-423

[35] Pielou EC. The measurement of diversity in different types of biological collections. Journal of Theoretical Biology. 1969;**13**:131-144

[36] Magurran AE. Measuring Biological Diversity. London, UK: Blackwell Publishing Limited; 2004. 260 p

[37] Martins VA, Silva F, Lazaro LML, et al. Response of benthic foraminifera to organic matter quantity and quality and bioavailable concentrations of metals in Aveiro Lagoon (Portugal). PLoS One. 2015;**10**(2):e0118077. DOI: 10.1371/journal.pone

[38] Martins MVA, Laut L, Duleba W, et al. Sediment quality and possible uses of dredged materials: The Ria de Aveiro lagoon mouth area (Portugal). Journal of Sedimentary Environments. 2017;**2**(2):149-166. DOI: 10.12957/jse.2017.30055

[39] Lopes JF, Dias JM, Cardoso AC, Silva CIV. The water quality of the Ria de Aveiro lagoon, Portugal: From the observations to the implementation of a numerical model. Marine Environmental Research. 2016;**60**(5):594-628. DOI: 10.1016/j.marenvres.2005.05.001

[40] Diz P, Francés G. Postmortem processes affecting benthic foraminiferal assemblages in the Ría de Vigo, Spain: Implications for paleoenvironmental studies. Journal of Foraminiferal Research. 2009;**39**(3):166-179

[41] Leorri E, Roland Gehrels W, Horton BP, Fatela F, Cearreta A. Distribution of foraminifera in salt marshes along the Atlantic coast of SW Europe: Tools to reconstruct past sea-level variations. Quaternary International. 2010;**221**:104-115. DOI: 10.1016/j.quaint.2009.10.033

[42] Debenay J-P, Guillou J-J. Ecological transitions indicated by foraminiferal assemblages in paralic environments. Estuaries and Coasts. 2002;**25**(6A):1107-1120. DOI: 10.1007/BF02692208

[43] Martins V, Isabel A, Carlos G, et al. Records of sedimentary dynamics in the continental shelf and upper slope between Aveiro-Espinho (N Portugal). Journal of Marine Systems. 2012;**96-97**:48-60. DOI: 10.1016/j.jmarsys.2012.02.001

[44] Martins MVA, Moreno JC, Miller PI, et al. a. Biocenoses of benthic foraminifera of the Aveiro Continental Shelf (Portugal): Influence of the upwelling events and other shelf processes. Journal of Sedimentary Environments. 2017;**2**(1):9-34. DOI: 10.12957/jse.2017.28041

[45] Frontalini F, Buosi C, da Pelo S, Coccioni R, Cherchi A, Bucci C. Benthic foraminifera as bioindicators of trace element pollution in the heavily contaminated Santa Gilla lagoon (Cagliari, Italy). Marine Pollution Bulletin. 2009;**58**:858-877. DOI: 10.1016/j.marpolbul. 2009.01.015

[46] Teodoro AC, Duleba W, Gubitoso S, Prada SM, Lamparelli CC, Bevilacqua JE. Analysis of foraminifera assemblages and sediment geochemical properties to characterise the environment near Araçá and Saco da Capela domestic sewage submarine outfalls of São Sebastião Channel, São Paulo State, Brazil. Marine Pollution Bulletin. 2010;**60**:536-553. DOI: 10.1016/j.marpolbul.2009.11.011

[47] Moreno J, Valente T, Moreno F, Fatela F, Guise L, Patinha C. Occurrence of calcareous foraminifera and calcite carbonate equilibrium conditions—A case study in Minho/Coura estuary (Northern Portugal). Hydrobiologia. 2007;**587**: 177-184. DOI: 10.1007/s10750-007-0677-7

[48] Murray JW, Alve E. Natural dissolution of modern shallow water benthic foraminifera: Taphonomic effects on the paleoecological record. Palaeogeography, Palaeoclimatology, Palaeoecology. 1999;**146**:195-209

[49] Sen Gupta BK, Machain-Castillo ML. Benthic foraminifera in oxygen-poor habitats. Marine Micropaleontology. 1993;**20**(3-4):183-201. DOI: 10.1016/0377-8398(93)90032-S

[50] Josefson AB, Widbom B. Differential response of benthic macrofauna and meiofauna to hypoxia in the Gullmar fiord basin. Marine Biology. 1988;**100**:31-40. DOI: 10.1007/BF00392952

[51] Armynot du Châtelet E, Degre D, Sauriau P-G, Debenay J-P. Distribution of living benthic foraminifera in relation with environmental variables within the Aiguillon cove (Atlantic coast, France): Improving knowledge for paleoecological interpretation. Bulletin de la Societe Geologique de France. 2009;**180**(2):131-144. DOI: 10.2113/gssgfbull.180.2.131

[52] Hottinger L, Reiss Z, Langer M. Spiral canals of some Elphidiidae. Micropaleontology. 2001;**47**(2):5-34 http://www.jstor.org/stable/1486160

[53] Duleba W, Teodoro AC, Debenay J-P, et al. Environmental impact of the largest petroleum terminal in SE Brazil: A multiproxy analysis based on sediment geochemistry and living benthic foraminifera. PLoS One. 2018;**13**(2):e0191446. DOI: 10.1371/journal.pone.0191446

[54] Castro-Filho BM, Miranda LB. Hydrographic properties in the São Sebastião Channel: Daily variations observed in March 1980. Revista Brasileira de Oceanografia. 1998;**46**:111-123

[55] Fortis RM, Ortiz JP, Lamparelli CC, Nieto R. Análise computacional comparativa da dispersão da pluma do efluente dos emissários submarinos do Tebar–Petrobrás. Revista Brasileira Recursos Hídricos. 2007;**12**:117-132

[56] Jackson RE, Reddy KJ. Trace element chemistry of coal bed natural gas produced water in the Powder River Basin, Wyoming. Environmental Science and Technology. 2007;**41**:5953-5959

[57] Walton WR. Techniques for recognition of living foraminifera.

Contribution of Cushman Foundation for Foraminifer Research. 1952;**3**:56-60

[58] Fatela F, Taborda R. Confidence limits of species proportions in microfossil assemblages. Marine Micropaleontology. 2002;**45**:169-174. DOI: 10.1016/S0377-8398(02)00021-X

[59] Murray JW. Foraminiferal assemblage formation in depositional sinks on the continental shelf west of Scotland. Journal of Foraminiferal Research. 2003;**33**:101-121

[60] Roberts DA, Johnston EL. Contaminants reduce the richness and evenness of marine communities: A review and meta-analysis. Environmental Pollution. 2009;**157**:1745-Johnston,1752

[61] Martins V, Jouanneau J-M, Weber O, Rocha F. Tracing the late Holocene evolution of the NW Iberian upwelling system. Marine Micropaleontology. 2006;**59**:35-55. DOI: 10.1016/j.marmicro.2005.12.002

[62] Bandy OL, Ingle JC, Resig JM. Foraminiferal trends, Hyperion outfall, California. Limnology and Oceanography. 1965;**10**:314-332

[63] Murray JW. Ecology and Applications of Benthic Foraminifera. New York, Melbourne: Cambridge; 2006. 426 pp. DOI: 10.1017/CBO9780511535529

[64] Viana AR, Faugères JC. Upper slope sand deposits: The example of Campos Basin, a latest Pleistocene-Holocene record of the interaction between alongslope and downslope currents. Geological Society, London, Special Publications. 1998;**129**:287-316

[65] Castro BM, Lorenzzetti JA, Silveira ICA, Miranda LB. Estrutura termohalina e circulação na região entre Cabo de São Tomé (RJ) e o Chuí (RS). In: Rossi-Wongtschowski CLDB, Madureira LS-P, editors. O Ambiente oceanográfico da plataforma continental e do talude na região Sudeste-Sul do Brasil. São Paulo: Editora da Universidade de São Paulo; 2006. pp. 11-120

[66] Ribeiro-Ferreira VP, Curbelo-Fernandez MP, Filgueiras VL, et al. Métodos empregados na avaliação do compartimento bentônico da Bacia de Campos. In: Falcão APC, Lavrado HP, editors. Ambiente Bentônico: Caracterização ambiental regional da Bacia de Campos, Atlântico Sudoeste. Rio de Janeiro: Elsevier; 2017. pp. 15-39. DOI: 10.1016/B978-85-352-7263-5.50002-3

[67] Loeblich AR, Tappan H. Foraminiferal genera and their classification. New York: van Nostrand Reinhold Company; 1988. pp. 1-2, 970 pp

[68] Ellis BF, Messina A. Catalogue of Foraminifera. New York: American Museum of Natural History; 1940

[69] Disaró ST, Aluizio R, Ribas E, et al. Foraminíferos bentônicos na plataforma continental da Bacia de Campos. In: Falcão APC, Lavrado HP, editors. Ambiente Bentônico: Caracterização ambiental regional da Bacia de Campos. Atlântico Sudoeste. Habitats. Vol. 3. Rio de Janeiro: Elsevier; 2017. pp. 111-144. DOI: 10.1016/B978-85-352-7263-5.50004-7

[70] Pianka ER. On r- and k-selection. The American Naturalist. 1970;**104**:592-597

[71] Warwick RM. A new method for detecting pollution effects on marine macrobenthic communities. Marine Biology. 1986;**92**:557-562

[72] Almeida AG, Kowsmann RO. Geomorfologia do talude continental e do Platô de São Paulo. In: Kowsmann RO, editor. Geologia e Geomorfologia: Caracterização

ambiental regional da Bacia de Campos, Atlântico Sudoeste. Série Habitats. Vol. 1. Rio de Janeiro: Elsevier. 2015. pp. 33-66. DOI: 10.1016/B978-85-352-6937-6.50010-0

[73] Sousa SHM, Passos RF, Fukumoto MM, et al. Mid-lower bathyal benthic foraminifera of the Campos Basin, Southeastern Brazilian margin: Biotopes and controlling ecological factors. Marine Micropaleontology. 2006;**61**:40-57. DOI: 10.1016/j.marmicro.2006.05.003

[74] Falcão APC, Curbelo-Fernandez MP, Borges ALN, Filgueiras VL, Kowsmann RO, Martins RP. Importância ecológica e econômica da Bacia de Campos: Ambiente transicional na margem continental do Oceano Atlântico Sudoeste. In: Curbelo-Fernandez MP, Braga AC, editoras. Ambiente Bentônico: Caracterização ambiental regional da Bacia de Campos, Atlântico Sudoeste. Habitats. Vol. 3. Rio de Janeiro: Elsevier; 2017. pp. 1-13. DOI: 10.1016/B978-85-352-7263-5.50001-1

[75] Yamashita C, Sousa SHM, Vicente TM, et al. Environmental controls on the distribution of living (stained) benthic foraminifera on the continental slope in the Campos Basin area (SW Atlantic). Journal of Marine Systems. 2018;**181**:37-52. DOI: 10.1016/j.jmarsys.2018.01.010

[76] Sousa SHM, Yamashita C, Nagai RH, et al. Foraminíferos bentônicos no talude continental, Platô de São Paulo e cânions da Bacia de Campos. In: Falcão APC, Lavrado HP, editors. Ambiente Bentônico: Caracterização ambiental regional da Bacia de Campos, Atlântico Sudoeste. Habitats. Vol. 3. Rio de Janeiro: Elsevier; 2017. pp. 111-144. DOI: 10.1016/B978-85-352-7263-5.50005-9

[77] Carreira RS, Araújo MP, Costa TLF, Ansari NF, Pires LCM. Lipid biomarkers in deep sea sediments from the Campos Basin, SE Brazilian continental margin. Organic Geochemistry. 2010;**41**:879-884